Critical Currents and Superconductivity

Ferromagnetism Coexistence in High-T_C Oxides

Critical Currents and Superconductivity

Ferromagnetism Coexistence in High-T_C Oxides

Samir Khene
Department of Physics, Faculty of Sciences
Badji Mokhtar University of Annaba
Annaba, Algeria

CRC Press
Taylor & Francis Group
Boca Raton London New York

CRC Press is an imprint of the
Taylor & Francis Group, an **informa** business

A SCIENCE PUBLISHERS BOOK

CRC Press
Taylor & Francis Group
6000 Broken Sound Parkway NW, Suite 300
Boca Raton, FL 33487-2742

First issued in paperback 2020

ISBN-13: 978-1-4987-7510-6 (hbk)
ISBN-13: 978-0-367-78299-3 (pbk)

Library of Congress Cataloging-in-Publication Data

Names: Khene, Samir, author.
Title: Critical currents and superconductivity : ferromagnetism coexistence in high-Tc oxides / Samir Khene.
Description: Boca Raton, FL : CRC Press, Taylor & Francis Group, [2016] | "2017 | Includes bibliographical references and index.
Identifiers: LCCN 2016007000| ISBN 9781498775106 (hardcover ; alk. paper) | ISBN 1498775101 (hardcover ; alk. paper)
Subjects: LCSH: Superconductivity. | Critical currents. | High temperature superconductors--Materials. | Copper oxide superconductors--Materials. | Ferromagnetism.
Classification: LCC QC611.92 .K44 2016 | DDC 537.6/23--dc23
LC record available at http://lccn.loc.gov/2016007000

Visit the Taylor & Francis Web site at
http://www.taylorandfrancis.com

and the CRC Press Web site at
http://www.crcpress.com

Preface

The field of superconductivity is constantly evolving. Very important discoveries were made since the beginning of the last century; some of them have even rewarded with Nobel Prizes.

In 1911, K.H. Onnes discovered that the electrical resistivity of many metals vanishes below certain very low critical temperatures (Nobel Prize).

In 1933, W. Meissner and R. Ochsenfelds showed that cooled to temperatures below its critical temperature, a superconductor expels the magnetic field.

In 1935, F. and H. London followed in 1950 by V.L. Ginzburg and L.D. Landau developed phenomenological theories which provided a better understanding of superconductivity (Nobel Prize).

Based on these models, A. Abrikosov presented in 1957 a theory of the mixed state of type-II superconductors, which stipulates that the magnetic flux penetrates in these materials in the form of vortices (Nobel Prize).

The same year, J. Bardeen, L.N. Cooper and J.R. Schrieffers elucidated the physical causes of the superconductivity phenomenon (Nobel Prize). In 1962, B.D. Josephson explained the tunneling junction behavior between the superconductors (Nobel Prize). Around the same time, the discovery of type-II superconductors which support very high magnetic fields (20 teslas) led to their intensive use for the generation of strong fields.

In 1986, J.G. Bednorz and K.A. Muller discovered superconductivity in a copper and lanthanum oxide doped with barium with a critical temperature of the order of 30K (Nobel Prize). This was the beginning of the high-T_c superconductors' era.

The highest critical temperature reached to date is 133K in a compound of the type $HgBaCa_{n-1}CunO_{2n+2+d}$ with $n = 3$, at ambient pressure. It reached 164K in the $HgBa_2Ca_2Cu_3O_x$ compound under high pressure. With out one well understands the physical mechanisms, responsible for the properties of these materials. However, the common feature of these new superconductors is the lamellar structure made up of elements with poor conductivity which are juxtaposed to copper-oxygen superconducting layers. This quasi-two-dimensional character induces in these compounds the anisotropy of all their superconducting properties. Indeed, the electrical conductivity is very high along the ab planes, whereas it is much lower in the perpendicular direction to them. The critical current is very large when the current circulates in these planes, and it is much lower in the perpendicular direction. The critical fields are higher in the direction of planes than in the perpendicular direction.

This book consists of six chapters. It begins by gathering key data for superconducting state and the fundamental properties of the conventional superconductors, followed by a recap of the basic theories of the superconductivity. It then discusses the differences introduced by the structural anisotropy on the Ginzburg-Landau approach and the Lawrence-Doniach model before addressing the dynamic of vortices and the ferromagnetism-superconductivity coexistence in high-Tc oxides, and provides an outline of the pinning phenomena of vortices in these materials, in particular the pinning of vortices by the spins.

This book elicits the methods to improve the properties of super-conducting materials for industrial applications. This optimization aims at obtaining critical temperatures and densities of critical current as high as possible. Whereas the primary objective concerns the basic mechanisms pushing the superconductivity towards high temperatures, the secondary objective is to achieve a better understanding of the vortices pinning. It is suitable for students of various streams of higher education and for the students in the doctoral program of all branches of the fundamental sciences and the engineering sciences. It will be beneficial to the experienced teachers and researchers. I thank S. Senoussi, D. Feinberg, B. Barbara and G. Fillion for all their help.

Prof. Dr. Samir Khene
Badji Mokhtar University of Annaba (Algeria)

Contents

Chapter 1

Superconducting State

1. History of the Superconductivity

The superconducting state is one of the most fascinating states of condensed matter. Scientific research within this domain has constantly evolved over the last century. During this period very important discoveries were made and some of these were awarded the Nobel Prize.

In 1911, three years after the first sucessful attempt at the liquefaction of helium. H.K. Onnes[1] discovered that the electrical resistance of mercury suddenly droped to zero for a certain critical temperature T_C below 4.2 kelvins. He called this new phenomenon "superconductivity". A year later, he observed that the current density values greater than a threshold value, called critical current density J_C, restored the resistance of mercury[2] and that this phenomenon also occured from a well-defined value of the applied field called a critical field H_C[3]. H.K. Onnes also discovered the superconductivity of tin and lead with critical temperatures of 3.7K and 7.2K, respectively[4]. In the years that followed, other pure superconducting elements were discovered at very low transition temperatures. The record for the critical temperatures of the pure elements is held by niobium with a transition temperature T_C of 9.2K.

In 1933, W. Meissner and R. Ochsenfeld[5] showed that a superconducting material expels the magnetic field when it is submitted to a weaker magnetic field and cooled to temperatures below its critical temperature. This remarkable phenomenon was called the Meissner effect.

In 1935, F. and H. London[6] proposed a macroscopic two-fluid model, normal fluid constituted of unpaired electrons and fluid formed by superconducting electrons (an idea originally introduced by C. Gorter and H.B.G. Casimir in 1934[7]). This model explained the Meissner effect and predicted the existence of a characteristic length of the superconducting state: the penetration depth called "penetration depth of London".

This was a certain depth measured which started at the surface of the superconductor where the magnetic flux persisted when the sample was in the Meissner state. Its amplitude exponentially decreased in the direction of the material center.

In 1950, F. London[8] suggested that the unusual properties of the superconductivity were the result of the occupation of a single quantum state called "condensate" (Pauli principle forbids two fermions, particles of spin 1/2, to occupy the same quantum state, but the bosons, particles of spin 1, naturally condenses in a same state). In the same year, V.L. Ginzburg and L.D. Landau[9] developed a phenomenological theory of the superconductivity which described the characteristic properties of most superconductors including those with high-T_C oxides (Nobel Prize). One of the greatest successes of this theory, which lies within the more general framework of the Landau theory[10] of phase transitions, is said to have predicted the existence of the type-II superconductors.

Based on the Ginzburg-Landau model, A. Abrikosov[11] presented in 1957 the theory of the mixed state of the type-II superconductors which stipulated that the magnetic flux penetrates the superconductor in the form of elementary entities called "vortices", each one carrying a quantum of flux (this theory was awarded the Nobel Prize). This quantification of the magnetic flux was actually predicted by F. London at an earlier date[8] which nevertheless had resulted in an elementary flux twice the size because London had considered the unpaired electrons.

In 1937, L.V. Shubnikov et al.[12] highlighted this new phase of the superconducting state. In 1964, D. Cribier et al.[13] proved the existence of the vortices lattice from the neutrons diffraction measurements, and in 1967, U. Essmann and H.T. Träuble[14] observed it with an electronic microscope using a decorative technique. In 1957, J. Bardeen, L.N. Cooper and J.R. Schrieffer[15] elucidated the physical causes of the superconductivity by offering a complete microscopic theory of this phenomenon where the fundamental assumption stipulated that superconductivity occurs when the attraction between two electrons via phonons prevailed over the usual Coulomb repulsion (Nobel Prize).

This idea that one originally owed to H. Fröhlich[16] was correctly exploited by L. Cooper[17]. This BCS theory perfectly described superconductivity in the conventional superconductors and constituted a required passage for researchers who are currently trying to establish the theory of the superconductivity of high-T_C oxides.

In 1962, B.D. Josephson[18] theoretically expected (and the experience confirmed it within the same year) that if two identical superconductors are separated by a very thin insulating region (not more about ten angstroms), the Cooper pairs can pass by tunneling effect through the junction without it dissociating as long as the applied voltage remains lower than a certain threshold voltage. Moreover, if one imposed a certain voltage, the current oscillated with a certain frequency which only depended on the fundamental constants of its physics (This theory was awarded the Nobel Prize). In addition to their own interest as a means of the study of the superconductivity, these junctions called "Josephson junctions" were the basis of numerous applications of superconductors in the electronic.

Around the same time, the discovery of the type II superconductors which supported very high magnetic fields (20 teslas) lead to their intensive use in the generation of strong fields.

Until early 1986, superconductivity was considered to be a low temperature phenomenon for the pure elements at an ambient pressure, transition temperatures ranging from 0.012K for tungsten to 9.2K for niobium and in the case of the alloys and compounds, a maximal critical temperature of the order of 23K for the Nb_3Ge compound. Thirty metallic elements had T_C of the order of one kelvin whereas the good conductors of electricity such as the noble metals (copper, silver and gold) and the alkali metals (sodium, potassium) were not superconductors even at very low temperatures. Further testing showed it was the same for magnetic metals (iron, cobalt and nickel) which showed no signs of superconductivity.

These superconductors henceforth refered to as "conventional superconductors" were divided into several classes which included polyvalent metals such as lead, tin, indium and their alloys, the transition alloys and compounds whose A15 type materials had the highest transition temperatures. The properties of these superconductors was well understood in the framework of the BCS theory.

Although the main phases present in ceramic oxides were discovered in the 1970's[19,20], scientists did not persevere in their efforts because the widespread idea within the scientific community at the time stated that superconductivity could not appear beyond 25K, proved to be a major psychological obstacle to any progress in this direction.

The critical temperatures of these oxides were very low because the density of their carriers was weak, with the exception of the compounds $LiTi_2O_4$ and $BaPbBiO_3$ for which the T_C was relatively high, of the order of 13K. The systematic research of superconductivity led by J.G. Bednorz and K.A. Muller[21] (I.B.M., Zurich) in the families of metallic oxides based on the copper or the nickel lead in 1986 to the discovery of superconductivity in an oxide of copper and lanthanum doped with barium with a critical temperature of the order of 30K (this was awarded the Nobel Prize). This was the beginning of the era of high-T_C superconductors.

The scientists then realized that it was possible to obtain critical temperatures of 20-40K for the whole family of $La_{2-x}M_xCuO_{4-y}$ with M = Ba, Sr, Ca and that by increasing the pressure, one can even exceed the 50K for M = Sr[22]. Understanding the importance of pressure to increase critical temperature, some researchers in Alabama and Houston, coordinated by K. Wu[23] and P. Chu[24], simulated a chemical pressure by replacing the atoms of lanthanum by atoms of yttrium and discovered in early 1987 the oxide of yttrium, barium and copper $Y_1Ba_2Cu_3O_{7-\delta}$ with a T_C of about 92K. This discovery was very important because for the first time in the world a superconducting transition was achieved at a temperature above the liquefaction temperature of nitrogen ($T_{liquid\ nitrogen}$ = 77K), liquid nitrogen is much cheaper than liquid helium. In addition, ceramic is easy to prepare by mixing, calcining and oxidation of its powder constituents. A year later, the bismuth compound, $Bi_2Sr_2CaCu_2Ox$[25,26], and the thallium compound $Tl_2Ba_2Ca_2Cu_3O_{10}$[27] were discovered with respective transition temperatures of 110 and 125K. The highest critical temperature reached to date is about 133K in the $HgBaCa_{n-1}CunO_{2n+2+\delta}$ compound with $n = 3$[28] at ambient pressure, it can reach 150K under high pressure[29,30]. Other researchers claim to have discovered some materials with higher critical temperatures, but none of these discoveries is reproducible or confirmed by other research laboratories. Although the physical mechanisms responsible for the properties of these materials remain unknown to us in the present, the common feature of this new generation of superconductors is their lamellar structure composed of low conducting elements which are juxtaposed superconducting layers of copper-oxygen. This quasi-two-dimensional character is at the root of the anisotropy of their superconducting properties.

2. Definition of a Superconducting Material

A material is said to be a superconductor if it has the following two remarkable properties:

i) $\rho = 0$ for all $T < T_C$: cooled below a certain very low temperature called a "critical temperature T_C", a superconducting material has a zero electrical resistivity (i.e. an infinite electric conductivity). Let us mention for example the total annulment of the mercury resistivity below 4.2K in a very short temperature interval, of the order of 0.05K[1].

ii) $\mathbf{B} = 0$ within the material: placed in a weak magnetic field, the superconductor expels the magnetic flux when it is cooled below T_C. This is the Meissner effect; it permits for the testing by very simple methods any materials, which are a candidate to superconductivity.

REMARKS

i) The annulment of the magnetic flux within the superconductor is not a consequence of its infinite electrical conductivity: these two phenomena are not linked and are two independent features of the superconducting state.

ii) Even cooled below its critical temperature T_C, a superconducting material loses its superconducting properties when it is subjected to a strong magnetic field. This leads us to define a certain internal field $B_C(T)$, the thermodynamic critical field, above of which the superconductivity is destroyed. This field depends on the temperature and the material; it vanishes at T_C.

iii) Crossed by an electrical current, a superconducting material cannot withstand very high current densities: there is a limit beyond which the material opposes a non-zero resistance to the current flow. This limit, denoted J_C, is called "critical current density". It conditions the use of the superconducting materials as conductors of electricity. It depends on the temperature and the magnetic field. In conclusion, a material keeps its superconducting properties only if its temperature is below T_C, the applied field lower than H_C (in the vacuum, one has $B = \mu_0 H$ where μ_0 is the magnetic permeability of the vacuum, it is worth $4\mu \times 10^{-7}$ MKSA) and the current density which traverses it is lower than J_C.

3. Meissner Effect

In a weak field, a superconductor cooled at temperatures below its critical temperature expels the lines of magnetic flux which initially crossed it. This remarkable phenomenon was discovered by W. Meissner and R. Ochsenfeld[5] in 1933 and is known as the Meissner effect. The material then behaves as a perfect diamagnetic material. In fact, the Meissner effect is not always achieved in volume. There are two distinct behaviors to the application of an external magnetic field giving rise to two types of superconductors, the type-I superconductors where the diamagnetism is perfect up until the value of the critical field B_C and above which the material again becomes normal and type-II superconductors where the diamagnetism is perfect until the first critical field B_{C1}, it becomes partial between B_{C1} and B_{C2} and disappears beyond the second critical field B_{C2}.

Table 1.1. Superconducting pure elements of the type I. T_C is the highest reported transition temperature and B_C is the thermodynamic critical field.

Formula	T_C(K)	B_C(T)	Reference
AL	1.20	0.01	[31–33]
Ga	1.083	0.0058	[34]
Re	2.4	0.03	[35]

All superconducting pure elements are of the type I except niobium and vanadium (Table. 1.1) whereas the alloys, the compounds and the high-T_C oxides are of the type II (Table. 1.2). Because of the very weak values of the critical field B_C, the type-I superconductors are not useful for the construction of coils for the superconducting magnets. On the contrary, the high values of the upper critical field B_{C2} of the type-II superconductors permit to use these superconductors for the manufacture of the magnets. A field B_{C2} of 41T was actually reached in an alloy of Nb, Al and Ge at the boiling temperature of the helium, of 51 teslas in the $Pb_1Mo_{5.1}S_6$ compound[36] and of the order of 140 T in $YBa_2Cu_3O_{7-\delta}$ thin films for H applied parallel to the **c**-axis of the film[37].

Table 1.2. Transition temperatures and upper critical fields of some type-II superconducting materials.

Formula	$T_c(K)$	$B_{c2}(T)$	Reference
In_2O_3	3.3	~ 3	[38]
MgB_2	39	74	[39]
Nb_3Sn	18.3	30	[40]

3.1. Type-I Superconductors

The magnetic field which destroys the superconductivity of the type-I material is referred to as "thermodynamic critical field B_C". Its variation as a function of temperature for several materials is nearly parabolic (see Fig. 1.1):

$$B_C = B_0 \left[1 - \left(\frac{T}{T_C} \right)^2 \right]$$ (1.1)

where B_0 represents the extrapolated value of B_C at $T = 0K$. In the international system, the magnetic field B (in teslas) which prevails in the material is given by the following relation:

$$\mathbf{B} = \mu_0 (\mathbf{H} + \mathbf{M})$$ (1.2)

where **M** is the magnetization of the sample in A/m, **H** the field of excitation in A/m and μ_0 the magnetic permeability of the vacuum. The Meissner state corresponds in the case where $\mathbf{B} = \mathbf{0}$ leading to $\mathbf{M} = -\mathbf{H}$: the diamagnetism is therefore perfect until the critical field B_C above which the material becomes normal, so $\mathbf{M} = \mathbf{0}$ (see Fig. 1.2).

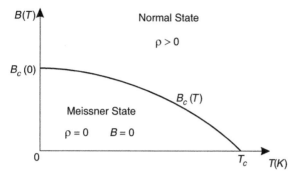

Figure 1.1. Phases diagram of a type-I superconductor.

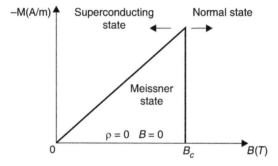

Figure 1.2. Magnetization as a function of the magnetic field of a type-I superconductor. The Meissner state is characterized by a perfect diamagnetism.

3.2. Type-II Superconductors

A type-II superconductor has two critical fields, the lower critical field B_{C1} and the upper critical field B_{C2}. For magnetic fields lower than B_{C1}, the type-II superconductor behaves like a type-I superconductor below B_C: it totally expels the magnetic flux (see Fig. 1.3).

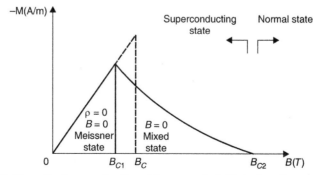

Figure 1.3. Magnetization as a function of the magnetic field of a type-II superconductor. In the mixed state, the Meissner effect is partial.

For fields higher than B_{C2}, the sample becomes normal. For fields between B_{C1} and B_{C2}, the magnetic flux partially penetrates into the sample in the form of very thin microscopic tubes called "vortices" (see Fig. 1.4). The material is then in a superconducting state called "mixed state"; the vortices are arranged according to a triangular configuration of Abrikosov[11] (see Fig. 1.5). Each vortex contains only one quantum of flux $\Phi_0 = 2.067 \times 10^{-15}$ Wb. Starting from its center, the vortex is formed of a normal core of radius ξ where the superconductivity is destroyed. This core is surrounded by a region of radius λ where the superconducting currents circumscribe the magnetic flux of the core (see Fig. 1.6). The field magnetic B is proportional to n, the number of vortices per m². This gradually increases when the magnetic field passes from B_{C1} to B_{C2} because one vortex contains only one quantum of flux. Close to B_{C1}, their number is small and the distance between the two neighboring vortices is about λ whereas for a field value near B_{C2}, they are so numerous that they touch each other, the distance between two neighboring vortices becomes ξ. The critical current J_C is the current which creates at the sample surface the field B_{C1} and not B_{C2} as one could think it because as soon as the first vortices appear in the sample (just above B_{C1}) they will be subjected to the Lorentz strength which forces them to move. This movement creates an electrical field and therefore a loss of energy, thus a resistivity. Therefore to make it pass high currents without loss of energy, it is necessary to prevent this movement of vortices by creating traps that the vortices cannot leave easily. The superconducting currents can then circulate in the superconducting regions between the vortices, and J_C will not be limited by B_{C1}.

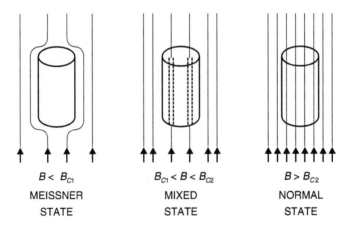

Figure 1.4. Different states in a type-II superconductor.

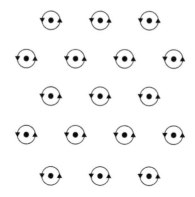

Figure 1.5. Triangular lattice of Abrikosov.

In the conventional type-II superconductors, the small non superconducting inclusions whose the size is about ξ, constitute an example of efficient traps for the vortices.

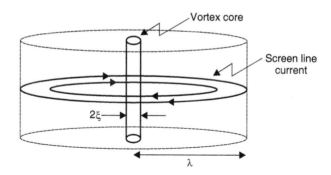

Figure 1.6. Structure of a vortex.

Such wires are used to manufacture the magnets which produce magnetic fields about ten teslas[36]. In the high-T_C oxides where ξ is very small, about ten angstroms, the mechanisms which govern the trapping of the vortices are not yet well-understood. The temperature dependence of the critical fields defines the phase diagram of a type II superconductor (see Fig. 1.7). This diagram is that of a perfect superconductor because the presence of defects by their trapping action on vortices can change it.

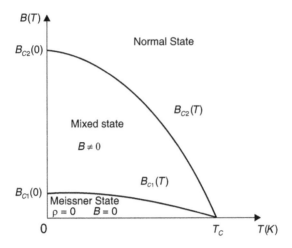

Figure 1.7. Phases digrams of a type-II superconductor, supposed perfect.

4. Description of the Superconducting State

4.1. Origin of Superconductivity

When a superconducting material is cooled below its critical temperature T_c, its electrical conductivity becomes infinite. This spectacular phenomenon can be understood only in the framework of the quantum description of the microscopic theory of BCS. Indeed, below T_c, the gas of normal electrons which characterized the normal state of the material before the transition turns into a quantic fluid of pairs of correlated electrons in the reciprocal space of the quantities of movement that is to say an electron with a given movement quantity and a given spin mates with another electron of an opposite movement quantity and an opposite spin in order to form a pair of zero quantity of movement called "pair of Cooper". The glue is ensured by the elastic waves of the crystalline lattice, known as phonons. The distance between the two electrons of the pair is not other than the coherence length ξ. It is usual to view the attraction between the two repulsive electrons by a naive description of the real space in the following manner: an electron interacts with the lattice of ions and distorts it by locally creating an excess of positive charge which has the effect of attracting a second electron. The second electron therefore interacts with the first through the lattice distortion.

4.2. Wave Function of the Superconducting State

The Cooper pair has twice the charge of the electron, i.e. $q = 2e$. While the unpaired electrons are fermions and obey to the statistics of Fermi-Dirac and to the principle of exclusion of Pauli which only authorizes one electron per quantum state, the Cooper pairs are quasi-bosons obeying to the statistics of Bose-Einstein which permit two different electrons in the same quantic state. In a metal, each electron has its own wave function. In the superconducting state, all the Cooper pairs are described by a single wave function:

$$\psi(\mathbf{r}) = \sqrt{n_S(\mathbf{r})} \, \exp[i\,\varphi(\mathbf{r})] \tag{1.3}$$

where $n_s(\mathbf{r})$ is the number of pairs (or superconducting electrons) with $\psi(\mathbf{r}).\psi^*(\mathbf{r}) = n_s(\mathbf{r})$, and $\varphi(\mathbf{r})$ the phase, macroscopic quantity proportional to the current which circulates in the superconductor.

4.3. Persistent Current

In order to verify that the resistivity of a superconductor was rigorously zero below T_c, H.K. Onnes[41] cools a superconducting material, starting from a temperature higher than the temperature T_C to a temperature lower than T_C; the field is then canceled inducing currents in the ring. For a ring of resistance R and of self-induction L, these currents decrease according to the law $I(t) = I(0) \exp(-Rt/L)$. By observing for example the torque exerted by the ring on another concentric ring crossed by a known current, it is possible to measure $I(t)$ with a very high degree of accuracy. In his experiments, H.K. Onnes used a coil of 700 meters of lead wire and did not detect any decrease of current during the 12 hours of experimentation. By his calculations, he arrived at a value of the resistance lower than $10^{-17} R_0$, R_0 is the resistance of the lead at ambient temperature (or lower than $10^{-15} R_0'$, with R_0' is the residual resistance at 0K, extrapolated as if the superconductivity did not appear). During a similar work done at the Massachusetts Institute of Technology (MIT), S.C. Collins[42] succeeded in 1956 to keep, for about two and a half years, a ring of lead in the superconducting state, crossed by a constant induced current of several hundreds of amperes. He deducted that the resistivity of the superconductor was less than 10^{-21} Ω.cm (for comparison, the low-temperature resistivity of pure copper is of the order of 10^{-10} Ω.cm). Then in 1963, J. File and R.G. Mills[43] studied the amortization of the currents in a superconducting solenoid by using very precise methods of Nuclear Magnetic Resonance (NMR) and deducted that the time of amortization of the superconducting current was superior than 100 000 years.

5. Electronic Specific Heat

At low temperatures, the variation of the electronic specific heat of a metal as a function of the temperature is described by a law of type:

$$C_{VN} \sim \gamma \, T \tag{1.4}$$

In superconductors, entropy which measures the disorder of a system considerably decreases during their cooling below T_C because electrons which were thermally excited in their normal state are arranged in the superconducting state. This variation of entropy is relatively weak. By cooling a superconductor in zero fields, its specific heat suddenly rises to T_C to its highest value and then smoothly decreases below its first value of the normal state (see Fig. 1.8). A thorough analysis of this phenomenon reveals that in the superconducting state, the linear electronic contribution to the specific heat is replaced by a term which quickly varies with temperature as:

$$C_{VS} \, (T < T_C) \sim \exp\left(-\frac{\Delta}{k_B T} \right) \tag{1.5}$$

where Δ is the superconducting gap at $T = 0K$ and $k_B \sim 1.38 \times 10^{-23}$ J/°K. This gap decreases when the temperature increases, it vanishes at $T = T_C$.

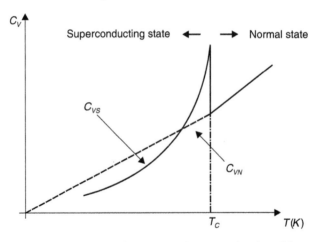

Figure 1.8. Electronic specific heat of a superconductor as a function of the temperature.

This variation characterizes the thermal behavior of a system whose excited levels are separated from their fundamental state by an energy $E_g = 2\Delta$ (see Fig. 1.9). BCS theory predicts the dependence of the value of the energy gap Δ at temperature T on the critical temperature T_C. The ratio between the value of the energy gap at zero temperature and the value of the superconducting transition temperature (expressed in energy units) takes the universal form[44]:

$$\Delta = 1.764 \times k_B \, T_C \tag{1.6}$$

Near the critical temperature, the previous relation asymptotes to[45]:

$$\Delta(T \to T_C) \approx 3.07 \times k_B T_C \times \sqrt{1 - \frac{T}{T_C}} \tag{1.7}$$

The gap in conventional superconductors is lower than the gap in high-T_C oxides. Moreover, in semiconductors, the gap in the energy spectrum corresponds to the energy difference between the valence band of the conduction band, whereas in the superconductors, E_g corresponds to the energy necessary to destroy the Cooper pair. Fig. 1.8 also shows that the transition in the absence of the magnetic field from the superconducting state to the normal state is of second order because there is no latent heat but a discontinuity in the electronic specific heat. In Table 1.3, one gives some pure elements values of $2\Delta/k_B T_C$.

Table.1.3. Measured values of $2\Delta/k_B T_C$ with an uncertainty of ± 0.1. The values of Δ at $T = 0K$ were obtained by tunneling effect (after R. Mersevey and B.B. Schwartz, 1969[46]).

Element	$2\Delta/k_B T_C$	Element	$2\Delta/k_B T_C$
Al	3.4	Pb	4.3
Cd	3.2	Sn	3.5
Hg (α)	4.6	Ta	3.6

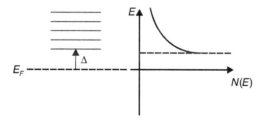

Figure 1.9. Density of states of a superconductor.

The energy gap (Δ) can be measured with high precision in a tunneling experiment (a process in quantum mechanics that allows an electron to escape from a metal without acquiring the energy required along the way according to the laws of classical physics).

6. Electromagnetic Absorption

The energy gap in a superconductor has a direct effect on the absorption of electromagnetic radiation. At low temperatures, at which a negligible fraction of the electrons are thermally excited to states above the gap, the superconductor can absorb energy only in a quantized amount that is at least twice the gap energy (at absolute zero, $2\Delta_0$). In the absorption process, a photon (a quantum of electromagnetic energy) is absorbed, and a Cooper pair is broken; both electrons in the pair become excited. The photon's energy E_g is related to its frequency by the Planck relation:

$$E_g = h\nu \tag{1.8}$$

in which h is Planck's constant (6.63×10^{-34} joule second) and ν the frequency in hertz (Hz). Therefore, the superconductor can absorb electromagnetic energy only for frequencies at least as large as $2\Delta_0/h$.

7. Isotopic Effect

Two different isotopes of the same superconductor have different critical temperatures T_C. The relationship between T_C and the atomic mass M of the isotope, which is valid for only a few materials, is:

$$T_C \sim M^{-\alpha} \tag{1.9}$$

where α is a number about 0.5 (see Table. 1.4). In mercury for example, the critical temperature T_C varies from 4.185K to 4.146K when the average mass of M varies from 199.5 to 203.4 atomic mass units[47-48].

Table 1.4. Experimental values of α in $M^{\alpha}T_C$ = Constant where M is the isotopic mass (adapted from J.W. Garland Jr., 1963[49]).

Material	α
Zn	0.45 ± 0.05
Cd	0.32 ± 0.07
Sn	0.47 ± 0.02
Hg	0.50 ± 0.03
Pb	0.49 ± 0.02
Mo_3Ir	0.33 ± 0.03

The variation of T_C with the isotopic mass indicates that the vibrations of the lattice and therefore the interactions between electrons and ions have something to do with the superconductivity.

8. Flux Quantification

The laws of quantum mechanics dictate that electrons have wave properties and that the properties of an electron can be summed up in what is called a wave function. If several wave functions are in phase (i.e., act in unison), they are said to be coherent.

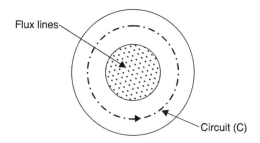

Flux lines

Circuit (C)

Figure 1.10. Flux lines through a superconducting ring.

The theory of superconductivity indicates that there is a single, coherent, quantum mechanical wave function that determines the behavior of all the superconducting electrons:

$$\psi(\mathbf{r}) = \sqrt{n_s(\mathbf{r})}\, \exp[i\,\varphi(\mathbf{r})] \tag{1.10}$$

in which $n_s(\mathbf{r})$ is the number of the cooper pairs (or the superconducting electrons) with $\psi(\mathbf{r}).\psi^*(\mathbf{r}) = n_s(\mathbf{r})$ and $\varphi(\mathbf{r})$ the phase. As a consequence, a direct relationship can be shown to exist between the velocity of these electrons and the magnetic flux (Φ) enclosed within any closed path inside the superconductor (Fig. 1.10). Indeed, in as much as the magnetic flux arises because of the motion of the electrons, the magnetic flux can be shown to be quantized; i.e., the intensity of this trapped flux can change only by units of Planck's constant divided by twice the electron charge:

$$\frac{h}{2\pi} \times 2\pi\, n = 2e\,\Phi \quad \Rightarrow \quad \Phi = n \times \frac{h}{2e} = n\,\Phi_0 \tag{1.11}$$

In which Φ_0 is the fundamental quantity ($\Phi_0 = 2.067 \times 10^{-15}$ Wb). First predicted by F. London[8] in 1950 with an elementary unit of flux 2 times bigger, this phenomenon of the quantification of the flux was discovered in 1961 by B.S. Deaver Jr., W.M. Fairbank[50], R. Doll and M. Nabauer[51] who demonstrated for the first time the existence of the Cooper pairs (the superconducting electrons of charge $q = 2e$), mentioned four years before in the BCS theory. Currently, there are devices that measure the tiny magnetic flux variation, which are very important for metrology and fundamental physics.

9. Josephson Effects

If two superconductors are separated by an insulating film (~nm) that forms a low-resistance junction between them, it is found that Cooper pairs can tunnel from one side of the junction to the other. This junction is known as a Josephson junction (Fig. 1.11). Thus, a flow of electrons, called the Josephson current, is generated and is intimately related to the phases of coherent quantum mechanical wave function for all superconducting electrons on two sides of the junction. It was predicted that several novel phenomena should be observable, and experiments have determinded them. These are collectively called Josephson effects[52].

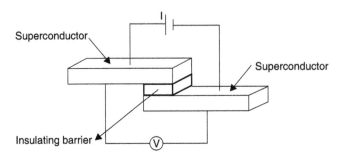

Figure 1.11. SIS Josepson junction.

9.1. DC Josephson Effect

If no voltage is applied to the junction, a current of Cooper pairs I flows through the junction up to a critical value I_c, which depends on the geometry, temperature and magnetic field. In exact terms, the current I is proportional to the surface of the junction and its value exponentially decreases with its thickness. The Cooper pairs current cannot exceed I_c. If a current $I > I_c$ is imposed, the difference of potential reaches V_c. Beyond this, the junction becomes resistive. The current will then be composed of normal carriers (dissociated pairs) and becomes almost linear in V (see Fig. 1.12).

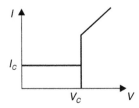

Figure 1.12. Electric current versus voltage in a SIS Josephson junction. If V exceeds V_c, the Cooper pairs dissociate and the junction becomes resistive.

Theoretical considerations connect V_c to Δ, the superconducting gap:

$$V_C = \frac{\pi\,\Delta}{2e} \tag{1.12}$$

9.2. AC Josephson Effect

A current $I > I_c$ can therefore move even for a zero voltage. But if one imposes a voltage V, the difference of phase γ will vary according to:

$$\frac{d\gamma}{dt} = \frac{2\pi}{h} \times Energie = \frac{2\pi}{h} \times 2eV \tag{1.13}$$

And the current will oscillate at a frequency:

$$\nu = \frac{2e}{h} \times V = \frac{V}{\Phi_0} \tag{1.14}$$

The ratio ν/V depends only on fundamental constants. It is equal to 484 MHz/µV. This is the AC Josephson Effect. This phenomenon is used in metrology for the accurate determination of the ratio h/e. Besides, application of a radiofrequency voltage (RF) superimposed on the continuous voltage can cause the passage of a continuous current through the junction. This fact is the basis of the working of the S.Q.U.I.D magnetometers. These two effects, confirmed experimentally, form the basis of the Josephson junctions whose applications relate to the electromagnetic waves detection, oscillators, electrical voltage standards, magnetically controlled logic circuits and dipoles[53].

REFERENCES

1   H.K. Onnes, The Disappearance of the Resistivity of Mercury, Comm. Leiden **122b** (27 May 1911).
2   H.K. Onnes, The Potential Difference Necessary for the Electric Current through Mercury below 4.19 K, Comm. Leiden **133a** (1913).
3   H.K. Onnes, The Appearance of Resistance in Superconductors which are Brought into a Magnetic Field, at a Threshold Value of Field, Comm. Leiden 139f (1914).
4   H.K. Onnes, The Sudden Disappearance of the Ordinary Resistance of Tin and the Superconductive State of Lead, Comm. Leiden **133d** (1913).
5   W. Meissner and R. Ochsenfeld, Naturwiss. **21**, 787 (1933).
6   F. London and H. London, Proc. Roy. Soc. **A149**, 71, London (1935).
7   C. Gorter and H.B.G. Casimir, Physica **1**, 306 (1934); C. Gorter and H.B.G. Casimir, Physik. Z. **35**, 963 (1934).
8   F. London, Superfluids, Vol. 1, Wiley, New York, 1950; reprint F. London, Superfluids, Vol. 1, Dover, New York, 1961.
9   V.L. Ginzburg and L.D. Landau, J.E.T.P., URSS **20**, 1064 (1950).
10 L.D. Landau, Zh. Eksp. Teor. Fiz. **7**, 19 (1937); **7**, 627 (1937); English translation in Collected Papers of L.D. Landau, edited by D. Ter Haar, Pergamon Press, Oxford, p. 193, 1965; see also L.D. Landau and E. Lifchitz, Statistical Physics, edition Mir, Moscow, Ch. XIV, 1967.
11 A.A. Abrikosov, Soviet Phys. JETP **5**, 1174 (1957); Fundamentals of the Theory of Metals, North-Holland, Amsterdam, 1988.

12 L.V. Shubnikov, V.I. Khotkevich, Yu.D. Shepelev and Yu.N. Riabinin, Zh. Eksperim. Teor. Fiz. 7, 221 (1937); Phys. Z. Sowjet **10**, 165 (1936).

13 D. Cribier, B. Jacrot, R.L. Madhav and B. Farnoux, Phys. Lett. **9**,106 (1964).

14 U. Essmann and H. Träuble, Phys. Lett. **24a**, 526 (1967).

15 J. Bardeen, L.N. Cooper and J.R. Schrieffer, Phys. Rev. **108**, 1175 (1957); **106**, 162 (1957).

16 H. Fröhlich, Phys. Rev. **79**, 845 (1950).

17 L.N. Cooper, Phys. Rev. **104**, 1189 (1956).

18 B.D. Josephson, Phys. Rev. Lett. **1**, 251 (1962); Rev. Mod. Phys. **36**, 216 (1964).

19 D.C. Johnston, H. Praksash, W.H. Zachariasen and R. Viswanathan, Mat. Res. Bull. **8**, 77 (1973).

20 A.W. Sleight, J.L. Gillson and P.E. Bierstedt, Solid St. Commun. **17**, 27 (1975).

21 J.G. Bednorz and K.A. Müller, Z. Phys. B **64**, 189 (1986).

22 H. Takagi, S. Uchida, K. Kitazawa and S. Tanaka, Jpn. J. Appl. Phys. **226**, 123 (1987).

23 M.K. Wu, J.R. Ashburn, C.J. Torng, P.H. Hor, R.L. Meng, L. Gao, Z.J. Huang, Y.G. Wang and C.W. Chu, Phys. Rev. Lett. **58**, 907 (1987).

24 C.W. Chu, P.H. Hor, R.L. Meng, L. Gao, Z.L. Huang and T.Q. Wang, Phys. Rev. Lett. **58**, 405 (1987).

25 H. Maeda, Y. Tanaka, M. Fukutomi and T. Asano, Jpn. J. Appl. Phys. Lett. **27**, L209 (1988).

26 C.W. Chu, J. Bechtold, L. Gao, P.H. Hor, Z.J. Huang, R.L. Meng, Y.Y. Sun, Y.Q. Wang and Y.Y. Xue, Phys. Rev. Lett. **60**, 941 (1988).

27 Z.Z. Sheng and A.M. Hermann, Nature **332**, 55 (1988).

28 S.N. Putilin, E.V. Antipov, O. Chmaissem and M. Marezio, Nature **362**, 226 (1993); S.N. Putilin, E.V. Antipov and M. Marezio, Physica C **212**, 266 (1993).

29 C.W. Chu, L. Gao, F. Chen, Z.J. Huang, R.L. Meng and Y.Y. Xue, Nature 365, 323 (1993).

30 M. Numez-Rugueiro, J.L. Tholence, E.V. Antipov, J.J. Capponi and M. Marezio, Science **262**, 97 (1993).

31 J.F. Cochran and D.E. Mapother, Physical Review **111** (1958).

32 B.T. Matthias and T.H. Geballe and V.B. Compton, Reviews of Modern Physics 35 (1963).

33 J. Eisenstein, Reviews of Modern Physics **26** (1954).

34 Kaxiras and Efthimios, Atomic and electronic structure of solids. Cambridge University Press. p. 283, 2003.

35 J.G. Daunt and T.S. Smith, Physical Review **88** (2), 309 (1952).

36 C. Kittel, Solid State Physics, Bordas, Paris, 1983.

37 M. Cyrot and D. Pavuna, Introduction to Superconductivity and Higt-T_C Materials, World Scientific, Singapore, 1992.

38 K. Makise, N. Kokubo, S. Takada, T. Yamaguti, S. Ogura, K. Yamada, B. Shinozaki, K. Yano, K. Inoue and H. Nakamura, Science and Technology of Advanced Materials 9 (4) (2008).

39 J. Nagamatsu, N. Nakagawa, T. Muranaka, Y. Zenitania and J. Akimitsu, (2001). Nature **410** (2001).

40 B.T. Matthias, T.H. Geballe, S. Gellera and E. Corenzwit Physical Review **95** (1954).

41 H.K. Onnes, The Imitation of an Ampere Molecular Current or of a Permanent Magnet by Means of a Superconductor, Comm., Leiden **104b** (1914).

42 S.C. Collins (not published) (quoted by P. Pugnat, Thesis, University Joseph Fourier-Grenoble I, 1995).

43 J. File and R.G. Mills, Phys. Rev. Lett. **10**, 93 (1963).

44 M. Tinkham, Introduction to Superconductivity, Dover Publications, p. 63, 1996.

45 M.J. Buckingham, Physical Review **101**, 1431 (1956).

46 R. Mersevey and B.B. Schawrtz, Superconductivity, R.D. Parks, ed. Dekker, New York, 1969.

47 E. Maxwell, Phys. Rev. **78**, 477 (1950).

48 Reynolds, Serin, Wright and Nesbitt, Phys. Rev. **78**, 487 (1950).

49 J.W. Garland Jr., Phys. Rev. Lett. **11**, 114 (1963); reviewed by V. Compton.

50 B.S. Deaver Jr. and W.M. Fairbank, Phys. Rev. Lett. **7**, 43 (1961).

51 R. Doll and M. Nabauer, Phys. Rev. Lett. **7**, 51 (1961)

52 B.D. Josephson, Phys. Rev. Lett., **1**, 251 (1962); Rev. Mod. Phys. **36**, 216 (1964).

53 P. Aigrain, Life Sciences, Proceedings, General Series, Volume 6, n°3, p. 211–229 (1989).

Chapter 2

Basic Models

1. London Model

In order to describe the Meissner effect[1], F. and H. London[2] modified a basic law of electrodynamics, the Ohm's law. Maxwell's equations always remain valid. To do this, they consider the superconductor as a two-fluid system, a fluid consisting of normal electrons and a fluid formed by superconducting electrons (idea originally introduced by Gorter and Casimir in 1934[3]). According to this two-fluid model, the first fluid consists of 'normal' electrons, of number density n_N, and these behave in exactly the same way as the free electrons in a normal metal. They are accelerated by an electric field E, but are frequently scattered by impurities and defects in the ion lattice and by thermal vibrations of the lattice. The scattering limits the speed of the electrons, and they attain a mean drift velocity:

$$v_N = - e\tau\, \mathbf{E} \tag{2.1}$$

Where τ is the mean time between scattering events for the electrons and m is the electron mass. The current density J_N due to flow of these electrons is:

$$\mathbf{J_N} = - n_N\, e\, \mathbf{v_N} = \frac{n_N e^2 \tau}{m}\mathbf{E} \tag{2.2}$$

The second fluid is formed by superconducting electrons with number density n_s. The superconducting electrons are not scattered by impurities, defects or thermal vibrations, so they are freely accelerated by an electric field. If the velocity of a superconducting electron is v_s, its equation of motion is:

$$m\frac{d\mathbf{v_s}}{dt} = - e\, \mathbf{E} \tag{2.3}$$

Combining this equation with the expression for the current density:

$$J_s = - n_s e v_s \qquad (2.4)$$

the equation becomes:

$$\frac{\partial J_s}{\partial t} = -n_s e \frac{\partial v_s}{\partial t} = \frac{n_s e^2}{m} E \qquad (2.5)$$

which is different to Equation 2.2. Scattering of the normal electrons leads to a constant current in a constant electric field, whereas the absence of scattering of the electrons in a superconductor means that the current density would increase progressively in a constant electric field. However, if one considers a constant current flowing in the superconductor, then:

$$\frac{\partial J_s}{\partial t} = 0 \qquad (2.6)$$

So:

$$E = 0 \qquad (2.7)$$

Therefore the normal current density must be zero, all of the steady current in a superconductor is carried by the superconducting electrons. Of course, with no electric field within the superconductor, there will be no potential difference across it, and so it has zero resistance. It was argued above that a material that just had the property of zero resistance, a perfect conductor rather than a superconductor, would maintain a constant magnetic field in its interior, and would not expel any field that was present when the material became superconducting. It shall now be shown how that conclusion follows from an application of Maxwell's equations to a perfect conductor. One can then see what additional assumptions are needed to account for the Meissner effect in a superconductor. If it is assumed that the electrons in a perfect conductor (or a proportion of them) are not scattered, and therefore the current density is governed by Equation 2.5. However, the subscript 'p' shall be used (for perfect conductor) here to indicate that it is not superconductor that is being dealt with. The interest is in the magnetic field in a perfect conductor, so Maxwell's equations will be applied to this situation. Faraday's law, valid in all situations, is given by:

$$\mathbf{curl}\, E = -\frac{\partial B}{\partial t} \qquad (2.8)$$

and if one substitutes E using Equation 2.5, one obtains:

$$\mathbf{curl}\, \frac{\partial J_p}{\partial t} = -\frac{\mu_0 n_p e^2}{m} \frac{\partial B}{\partial t} \qquad (2.9)$$

Looking now at the Ampère-Maxwell law:

$$\text{curl } \mathbf{H} = \mathbf{J}_F + \frac{\partial \mathbf{D}}{\partial t} \qquad (2.10)$$

it shall be assumed that our perfect conductor is either weakly diamagnetic or weakly paramagnetic, so that $\mu \simeq 1$ and $\mathbf{H} \simeq \mathbf{B}/\mu_0$ are very good approximations. Maxwell's term, $\partial \mathbf{D}/\partial t$ shall also be omitted, since this is negligible for the static, or slowly-varying, fields that shall be considered. With these approximations, the Ampère-Maxwell law simplifies to Ampère's law:

$$\text{curl } \mathbf{B} = \mu_0 \mathbf{J}_P \qquad (2.11)$$

where use of the subscript 'p' for the current density reminds us that the free current $\mathbf{J}_F$ is carried by the perfectly-conducting electrons. One now uses this expression to eliminate $\mathbf{J}_P$ from Equation 2.9:

$$\text{curl}\left(\text{curl}\frac{\partial \mathbf{B}}{\partial t}\right) = -\frac{\mu_0 n_p e^2}{m}\frac{\partial \mathbf{B}}{\partial t} \qquad (2.12)$$

A standard vector identity from inside the back cover to rewrite the left-hand side of this equation:

$$\text{curl}\left(\text{curl}\frac{\partial \mathbf{B}}{\partial t}\right) = \text{grad}\left(div\frac{\partial \mathbf{B}}{\partial t}\right) - \nabla^2\left(\frac{\partial \mathbf{B}}{\partial t}\right) \qquad (2.13)$$

The no-monopole law, div $\mathbf{B} = 0$, means that the first term on the right-hand side of this equation is zero, so Equation 2.12 can be rewritten as:

$$\nabla^2\left(\frac{\partial \mathbf{B}}{\partial t}\right) = \frac{\mu_0 n_p e^2}{m}\frac{\partial \mathbf{B}}{\partial t} \qquad (2.14)$$

This equation determines how $\partial \mathbf{B}/\partial t$ varies in a perfect conductor. One shall look for the solution of Equation 2.14. For the simple geometry shown in Fig. 2.1, a conductor has a boundary corresponding to the plane $z = 0$, and occupies the region $z > 0$, with a uniform field outside the conductor given by $\mathbf{B}_0 = B_0\,\mathbf{e}_x$. The uniform external field in the x-direction means that the field inside the conductor will also be in the x-direction, and its strength will depend only on z. So, Equation 2.14 reduces to the one-dimensional form:

$$\frac{\partial^2}{\partial z^2}\left(\frac{\partial B_x(z,t)}{\partial t}\right) = \frac{1}{\lambda_p^2}\frac{\partial B_x(z,t)}{\partial t} \qquad (2.15)$$

where the equation has been simplified, for reasons that will soon become clear, by writing:

$$\frac{\mu_0 n_p e^2}{m} = \frac{1}{\lambda_p^2} \tag{2.16}$$

The general solution of this equation is:

$$\frac{\partial B_x(z,t)}{\partial t} = a(t)e^{-x/\lambda_p} + b(t)e^{+x/\lambda_p} \tag{2.17}$$

Where a and b are independent of the position. The second term on the right-hand side corresponds to a rate of change of field strength that continues to increase exponentially with distance from the boundary; since this is unphysical, one sets $b = 0$. The boundary condition for the field parallel to the boundary is that $H_{//}$ is continuous, and since it is being assumed that $\mu \simeq 1$ in both the air and the conductor, this is equivalent to $B_{//}$ being the same on either side of the boundary at all times. This means that $\partial B/\partial t$ is the same on either side of the boundary, so:

$$a = \frac{\partial B_0}{\partial t} \tag{2.18}$$

and the field within the perfect conductor satisfies the equation:

$$\frac{\partial B_x(z,t)}{\partial t} = \frac{\partial B_0(t)}{\partial t} e^{-z/\lambda_p} \tag{2.19}$$

This indicates that any changes in the external magnetic field are attenuated exponentially with distance below the surface of the perfect conductor. If the distance λ_p is very small, then the field will not change within the bulk of the perfect conductor. Note that this does not mean the magnetic field must be expelled: flux expulsion requires $\mathbf{B} = 0$, rather than just $\partial B/\partial t = 0$. So how does one modify the description that has been given of a perfect conductor so that it describes a superconductor and leads to a prediction that $\mathbf{B} = 0$? In order to explain the Meissner effect, the London brothers proposed that in a superconductor, Equation 2.9 is replaced by the more restrictive relationship:

$$\mathbf{curl\ J_s} = -\frac{\mu_0 n_p e^2}{m} \mathbf{B} \tag{2.20}$$

This equation, and Equation 2.5 which relates the rate of change of current to the electric field, are known as the London equations. It is important to note that these equations are not an explanation of superconductivity.

They were introduced as a restriction on Maxwell's equations so that the behaviour of superconductors deduced from the equations was consistent with experimental observations and in particular with the Meissner effect. Their status is somewhat similar to Ohm's law, which is a useful description of the behaviour of many normal metals, but which does not provide any explanation for the conduction process at the microscopic level. To demonstrate how the London equations lead to the Meissner effect, one proceeds in the same way as for the perfect conductor. First we use Ampère's law, curl $\mathbf{B} = \mu_0 \mathbf{J}_s$, by substituting $\mathbf{J}_s$ in Equation 2.15 and obtain:

$$\mathbf{curl\ (curl\ J_s)} = -\frac{\mu_0 n_{pc} e^2}{m}\mathbf{B} = -\frac{1}{\lambda^2}\mathbf{B} \qquad (2.21)$$

Where:

$$\lambda = \sqrt{\frac{m}{\mu_0 n_p e^2}} \qquad (2.22)$$

But $\mathbf{curl(curl\ B)} = \mathbf{grad}\ (\text{div } \mathbf{B}) - \nabla^2 \mathbf{B} = -\nabla^2\mathbf{B}$, since div $\mathbf{B} = 0$. So:

$$\nabla^2\mathbf{B} = \frac{1}{\lambda^2}\mathbf{B} \qquad (2.23)$$

This equation is similar to Equation 2.14, but $\partial B/\partial t$ has been replaced by $\mathbf{B}$. The important point to note about this equation is that the only solution that corresponds to a spatially uniform field (for which $\nabla^2\mathbf{B} = 0$) is the field that is identically zero everywhere. If $\mathbf{B}$ was not equal to zero, then $\nabla^2\mathbf{B}$ would not be zero, so $\mathbf{B}$ would depend on the position. If again the simple one-dimensional geometry shown in Fig. 2.1 is considered, then the solution of Equation 2.19 is obtained by simply replacing the partial time derivatives of the fields in the solution for the perfect conductor (Equation 2.19) by the fields themselves, that is:

$$B_x(z) = B_0\ e^{-\frac{z}{\lambda}} \qquad (2.24)$$

Equation 2.24 shows that the magnetic field vanishes inside the material (see Fig. 2.1). The length λ which measures the depth of penetration of the magnetic field inside the superconductor is called "penetration depth of London". In the pure elements, λ is of the order of 10^{-6} cm, that is to say larger than the interatomic distance (see Table 2.1). It is clear that to have a zero field inside the material a supercurrent must circulate at a depth λ counted starting from its surface creating an opposite field to the applied field.

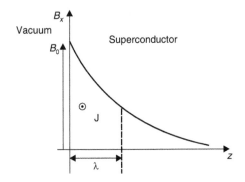

Figure 2.1. Penetration of the magnetic field inside the superconductor.

Table 2.1. London penetration depth at the absolute zero, of some pure elements (adapted from R. Meservey and B.B. Schwartz, 1969[4]).

Material	λ (10^{-6} cm)
Sn	3.4
Al	1.6
Pb	3.7
Cd	11.0
Nb	3.9

The proposed equations of the brothers Fritz and Heinz London in 1935 are thus consistent with the Meissner effect and can be used with Maxwell's equations to predict how the magnetic field and surface current vary with distance from the surface of a superconductor.

ADDITIONAL REMARK

It is pointed out that the magnetic field **B**, expressed in the international system in tesla is the field which reigns inside the superconducting material. The excitation field **H**, expressed in the same system in A/m is the field created by an external source to the sample and which is applied to it. In the material, these two fields are linked by the following equation:

$$\mathbf{B} = \mu_0 \, (\mathbf{H} + \mathbf{M}) \tag{2.25}$$

where **M** is the magnetization expressed in A/m and μ_0 the permeability of the vacuum. In the vacuum, these two quantities are proportional:

$$\mathbf{B} = \mu_0 \mathbf{H} \tag{2.26}$$

2. Phenomenological Theory of Ginzburg-Landau

The model of London does not take into account the variation in the number of the superconducting electrons n_s with the temperature, it does not either join this number to the applied field or to the current from where the need to have a more general theory that joins n_s to the external parameters. In 1950, V.L. Ginzburg and L.D. Landau[6] presented a phenomenological theory of superconductivity based on the theory of the phases transitions of the second order worked out by L.D. Landau[7] in 1937. Let us recall that in thermodynamics, a transition of phase is called of the second order if the entropy and the volume are continuous functions at the transition. In their theory, V.L. Ginzburg and L.D. Landau deduce two equations which permit calculation of the field distribution and the variation of the number of the superconducting electrons as a function of the temperature. For that reason, they introduce two unknown parameters, α and β which can be replaced by two of the three measurable quantities, namely the depth of penetration λ, the coherence length ξ and the thermodynamic critical field H_c. One of the greatest successes of this theory is to have predicted the existence of the superconductors of type II. The basic idea of this approach is to consider that the superconductor contains superconducting electrons of density n_s and normal electrons of density $n - n_s$, where n is the total density of the electrons within the material. Instead of using n_s to describe the superconducting electrons, V.L. Ginzburg and L.D. Landau chose a kind of wave function for the superconducting particle of mass m and charge q. Historically, the charge q was taken equal to e because in 1952 the concept of Cooper pairs was not yet known; This is only after the development of the BCS theory in 1957 that the charge q was replaced by $2e$. This function of wave, named "parameter of order", is a scalar number which is written:

$$\psi(\mathbf{r}) = |\psi(\mathbf{r})|\, e^{i\varphi(\mathbf{r})} \tag{2.27}$$

It has the following properties:
  i) Its modulus $|\psi^*\psi|$ represents the density of the superconducting electrons n_s.
  ii) Its phase $\varphi(\mathbf{r})$ is related to the superconducting current which circulates in the superconductor for $T < T_c$.
  iii) $\psi \neq 0$ in the superconducting state and $\psi = 0$ in the normal state; this parameter vanishes continuously at the transition.

Assuming smallness of $|\psi|$ and smallness of its gradients, the free energy has the form of a field theory:

$$F_s = F_N + \alpha|\psi|^2 + \frac{\beta}{2}|\psi|^4 + \frac{1}{2m}\left|\left(-\hbar\nabla - 2e\mathbf{A}\right)\psi\right|^2 + \frac{|\mathbf{B}|^2}{2\mu_0} \tag{2.28}$$

In which S and N are the respective indices of the superconducting state and the normal state and $\hbar$ represents Planck's constant ($\hbar = h/2\pi = 1.05 \times 10^{-34}$ MKSA), α and β in the initial argument were treated as phenomenological parameters, m is an effective mass, e is the charge of an electron, $\mathbf{A}$ is the magnetic vector potential, and $\mathbf{B} = \nabla \times \mathbf{A}$ is the magnetic field. By minimizing the free energy with respect to variations in the order parameter ψ and the vector potential $\mathbf{A}$, one arrives at two coupled equations called "Ginzburg-Landau equations":

$$\alpha\psi + \beta |\psi|^2 \psi + \frac{1}{2m} (i\hbar\nabla - 2e\,\mathbf{A})^2 \psi = 0 \qquad (2.29)$$

$$\nabla \times \mathbf{B} = \mu_0 \mathbf{J}; \quad \mathbf{J} = \frac{2e}{m} Re\,[\psi^* (-i\hbar\nabla - 2e\,\mathbf{A})\psi] \qquad (2.30)$$

where $\mathbf{J}$ denotes the dissipation-less electric current density and Re the real part. Equation 2.29 gives the parameter of order whereas Equation 2.30 describes the superconducting current $\mathbf{J}$. These two equations allow calculation of the characteristic quantities of the superconductor λ, ξ and H_C. Let us consider a homogeneous superconductor where there is no superconducting current, the equation for ψ simplifies to:

$$\alpha\psi + \beta|\psi|^2 \psi = 0 \qquad (2.31)$$

This equation has a trivial solution: $\psi = 0$. This corresponds to the normal state of the superconductor, that is for temperatures above the superconducting transition temperature, $T > T_C$. Below the superconducting transition temperature, the above equation is expected to have a non-trivial solution (that is $\psi \neq 0$). Under this assumption the equation above can be rearranged into:

$$|\psi_0|^2 = -\frac{\alpha}{\beta} \qquad (2.32)$$

No solution is possible except for $T < T_C$ where $\alpha = a\,(T - T_C)$ is negative, the system of minimal free energy is then in the superconducting state. If α is positive, the order parameter vanishes and the system switches to the normal state (see Fig. 2.2).

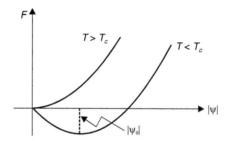

Figure 2.2. Free energy as a function of $|\psi|$. If α is positive, the system is in the normal state ($T > T_c$). On the other hand, if α is negative, the minimum of the free energy is obtained for a certain value $|\psi_0|$, the system is then in the superconducting state.

Ginzburg–Landau equations predicted two new characteristic lengths in a superconductor. One of them is termed coherence length, ξ. It represents the length on which the parameter of order varies when a disturbance is introduced at the point **r**. For $T > T_C$ (normal phase), it is given by:

$$\xi(T) = \sqrt{\frac{\hbar^2}{2m|\alpha|}} \tag{2.33}$$

while for $T < T_C$ (superconducting phase), where it is more relevant, it is given by:

$$\xi(T) = \sqrt{\frac{\hbar^2}{4m|\alpha|}} \tag{2.34}$$

The second one is the penetration depth, λ. It was previously introduced by the London brothers in their theory. Expressed in terms of the parameters of Ginzburg-Landau model, it is:

$$\lambda = \sqrt{\frac{m}{4\mu_0 e^2 |\psi_0|^2}} \tag{2.35}$$

where ψ_0 is the equilibrium value of the order parameter in the absence of an electromagnetic field. When $T \to T_C$, $|\psi_0| \to 0$, therefore $\alpha \to 0$ and $\xi \to \infty$: ξ, as besides λ, diverges at T_C.

As the two characteristic lengths ξ and λ diverge at T_C as $|\alpha|^{-1/2}$, their ratio κ, called "parameter of Ginzburg-Landau", does not depend on the temperature. It is given by:

$$\kappa = \frac{\lambda}{\xi} \tag{2.36}$$

This parameter is a complete phenomenological characteristic of a given superconductor.

Ginzburg-Landau theory[10]:

i) Gives a relationship which connects the parameters α and β to the thermodynamic critical field:

$$\mu_0 H_C^2 = \frac{\alpha^2}{\beta} \tag{2.37}$$

ii) Expresses α and β as a function of the measurable parameters by:

$$\alpha = - \frac{4e^2}{m} \mu_0^2 H_C^2 \lambda^2 \tag{2.38}$$

$$\beta = \left(\frac{4e^2}{m} \right)^2 \mu_0^3 H_C^2 \lambda^4 \tag{2.39}$$

iii) Gives an important relation between the measurable characteristic quantities H_C, λ and ξ:

$$H_C(T)\, \lambda(T)\, \xi(T) = \frac{\hbar}{2e\mu_0 \sqrt{2}} = \frac{\Phi_0}{2\pi\mu_0\sqrt{2}} = C^{te} \tag{2.40}$$

iv) Gives the upper critical field B_{C2}:

$$B_{C2}(T) = \frac{\Phi_0}{2\pi} \frac{1}{\xi^2(T)} = \mu_0 H_{C2} = \mu_0 \kappa \sqrt{2} H_C \tag{2.41}$$

Equation (2.39) permits to distinguish the two following situations:

- If $\kappa < 1/\sqrt{2}$, one has $B_{C2} < B_C$. By decreasing the field, the superconducting state appears at and below B_C with a total expulsion of the magnetic flux, the superconductor is of the type I.
- If $\kappa > 1/\sqrt{2}$, one has $B_{C2} > B_C$. The superconducting state appears at and below B_{C2}. As the expulsion of the flux is not complete, the superconductor is thus of the type II.

v) Gives the lower critical field B_{C1}:

$$B_{C1} = \frac{\Phi_0}{4\pi \lambda^2} \ln\frac{\lambda}{\xi} \tag{2.42}$$

vi) and a relation between B_{C1} and B_{C2}:

$$B_{C1}\, B_{C2} = B_C^2 \ln\kappa \tag{2.43}$$

As the value of B_c does not greatly vary in the type II superconductors, between 0.1 and 1T, a high value of B_{c2} leads to a very low value of B_{c1}: this is the case of high-T_c oxides where B_{c2} is about 100T whereas B_{c1} turns around 0.01T[10].

When studying superconductivity in Ginzburg-Landau theory in strong magnetic fields, one encounters three critical values of the magnetic field strength. The first critical field B_{c1} is where a vortex appears. At the second critical field, denoted B_{c2}, superconductivity becomes essentially restricted to the boundary and is weak in the interior. At the third critical field, B_{c3}, superconductivity disappears altogether. The third critical field is given as a function of the second critical field by:

$$B_{C3} = \frac{1}{0.59} \times B_{C2} = 1.69 \times B_{C2} \qquad (2.44)$$

This means that the surface of the sample parallel to the magnetic field B between B_{c2} and B_{c3} presents a superconducting layer of thickness of the order of ξ. Of course, if one reduces the applied field below B_{c2}, the superconductivity extends to the whole volume of the sample. The existence of a third line in the phase diagram of a type II superconductor, $B_{c3}(T)$ (see Fig. 2.3), can explain some facts of the experiment. Indeed, for an applied field parallel to the sample surface, the resistivity measurements give B_{c3} below which the resistivity vanishes and the superconducting current circulates in the superficial layer. Now, if measurements of magnetization are made, B_{c2} is obtained because this experiment is sensitive to the massive state of the sample. The B_{c3} field is not specific to type II superconductor, it has also been observed in some superconductors of type I.

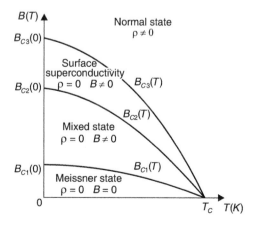

Figure 2.3. Variation as a function of the temperature of the three critical fields of a type II superconductor, of finite size.

3. BCS theory

3.1. *Main Results of the Theory*

The microscopic theory of the superconductivity was established in 1957 by J. Bardeen, L.N. Cooper and J.R. Schrieffer[12]. This theory justifies the phenomenological Ginzburg-Landau approach. It aptly describes the physical properties of the conventional superconductors and constitutes a theory of reference to interpret some experimental facts of high-T_C superconductors. Its main results can be summarized as follows[13]:

i) An attractive interaction between electrons may lead to a fundamental state of the electronic system separated from the excited states by a band gap. The critical field, the thermal properties and most of the electromagnetic properties are the result of this attraction.

ii) The electron-ion-electron interaction is attractive; it leads to a band gap. This interaction acts as follows: an electron interacts with the lattice of ions and locally deforms it creating an excess of positive charge which attracts a second electron; the second electron therefore interacts with the first electron via the lattice deformation.

iii) The characteristic lengths λ and ξ of the superconductor are the natural results of the fundamental state of BCS. The London equation is obtained for fields which slowly vary in the space and the Meissner effect, central phenomenon of superconductivity, is naturally obtained.

iv) For $U\,N(E_F) \ll 1$, where U is the electron-lattice attractive interaction and $N(E_F)$ the density of states at the Fermi level, the BCS theory predicts a variation of the transition temperature of the type:

$$T_C = 1.14 \times \Theta_D \, \exp\left[-\frac{1}{U\,N(E_F)}\right] \qquad (2.45)$$

where Θ_D is the Debye temperature. This result is qualitatively verified by experiment.

v) The quantization of the magnetic flux in a superconducting ring of unit of charge $2e$ rather than that e is a consequence of the quantum theory of BCS[14-16].

3.2. *Superconducting Gap*

BCS theory of superconductivity successfully describes the measured properties of type I superconductors. It envisions resistance-free conduction of coupled pairs of electrons called Cooper pairs. This theory is remarkable enough that it is interesting to look at the chain of ideas which led to it. The principal step toward a theory of superconductivity was the realization that there must be a band gap separating the charge carriers from the state of normal conduction. A band gap was implied by the very fact that the resistance is precisely zero. If charge carriers can move through a crystal lattice without interacting at all, it must be because their energies are quantized such that they do not have any available energy levels within reach of the energies of interaction with the lattice. A band gap is suggested by specific heats of materials like vanadium. The fact that there is an exponentially increasing specific heat as the temperature approaches the critical temperature from below implies that thermal energy is being used to bridge some kind of gap in energy. As the temperature increases, there is an exponential increase in the number of particles which would have enough energy to cross the gap. The fundamental state of a gas of Fermi electrons without interaction is the filled sphere of Fermi (Fig. 2.4).

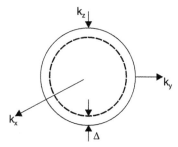

Figure 2.4. Spherical Fermi surface.

It is possible with this normal state to form an excited state by simply taking an electron from the Fermi surface and putting it just above this surface. The fundamental state of BCS is formed of a mixture of empty and filled states above and below the Fermi energy E_F. The filled states are occupied by pairs: if a state of wave vector $\mathbf{k}$ and of spin upwardly directed is occupied, the state of wave vector $-\mathbf{k}$ and of spin downwards directed is also occupied. If $\mathbf{k}'\uparrow$ is empty then $-\mathbf{k}'\downarrow$ is also empty. These pairs are called "Cooper pairs"; they have a null spin and many characteristics of bosons. By comparing the state of Fermi with the state of BCS, it seems that the energy of the superconducting state is higher than that of the normal state because the presence of occupied states above E_F increases the kinetic energy of the BCS state. In fact, the attractive potential energy of BCS level lowers the total energy of the system relative to the Fermi level. Δ, a fundamental quantity introduced by J. Bardeen, L.N. Cooper and J.R. Schrieffer, is called "superconducting gap". The superconducting gap depends on the temperature and obeys the following relationship[10]:

$$\frac{1}{UN(E_F)} = \int_0^{\omega_D} d\xi (\xi^2 + \Delta^2)^{-1/2}$$
$$\tanh\left[\frac{1}{2k_B T} (\xi^2 + \Delta^2)^{1/2}\right]$$

(2.46)

For $T = T_C$, one has $\Delta(T_C) = 0$: the energies of the normal state and the superconducting state are equal. An important equation of BCS connects Δ to T_C:

$$\frac{2\Delta}{k_B T_C} = 3.52$$

(2.47)

In general, the value of this ratio, measured by several methods, varies between 2.5 and 4.5 (see Table 2.2) and is close to 3.5 for many elements. The measured bandgap in type I superconductors is one of the pieces of experimental evidence which supports the BCS theory. Some materials (e.g. lead, mercury) produce experimental results which deviate substantially from the BCS results. These materials are more appropriately described by strong-coupling theory where the coupling ratio $\Delta/k_B T_C$ is greater than the BCS prediction of 1.76. Under certain conditions superconductivity can occur without an energy gap in some materials. Tunneling experiments on superconductors with specific concentrations of paramagnetic impurities show this to be possible. Theories exist which explain such anomalies, and the nature of the gap is a vital property to be considered in any theory describing superconductivity in the high-T_C compounds. The energy gain between the normal state E_N and the superconducting state E_S is given by:

$$E_N - E_S = \frac{1}{2} N(E_F) \Delta^2(T) = \frac{1}{2} \mu_0 H_C^2(T)$$

(2.48)

Near T_C, one has an approximate formula for $\Delta(T)$, valid for $T \le T_C$:

$$\frac{\Delta(T)}{\Delta} = 1.74 \times \left(1 - \frac{T}{T_C}\right)^{1/2}$$

(2.49)

Table 2.2. Experimental values of the ratio $2\Delta/k_B T_C$ at T = 0°K of some selected superconductors (adapted from B. Bleaney and B.I. Bleaney, 1976[9]).

Material	$2\Delta/k_B T_C$ (by tunneling effect)	$2\Delta/k_B T_C$ (by electromagnetic absorption)
Aluminum	2.5 - 4.2	3.37
Tin	2.8 - 4.1	3.5
Lead	4.3 - 4.4	4.0 - 4.4

In the fundamental state, all electrons are arranged by pairs. By breaking these pairs, one introduces excitations. This situation is different from that of metal where the excitations are produced by taking an electron below the Fermi level to fill a state which is above. To create an excitation of wave vector **k** in a superconductor costs energy. Contrary to the normal state, there is a gap in the excitation spectrum which depends on the temperature and becomes zero at T_C. The fundamental state is destabilized, its energy changes by a minimal quantity. For $T \neq 0$, one has Cooper pairs and single electrons. By not taking into account the effects of coherence, this situation appears as a simple two-fluid model where the superconducting electrons are the pairs and the normal electrons are the excitations formed of single electrons.

3.3. Comparison between Ginzburg-Landau Theory and BCS Theory

L.P. Gorkov established by Green function methods that the Ginzburg-Landau theory could be obtained from the BCS theory with some approximations. Gorkov's calculation also allows to show that $e^* = 2e$, $m^* = 2m$, that is to say, the "condensed electrons" of Ginzburg and Landau are actually pairs of electrons. The difference between these two theories is that the BCS theory is derived bottom-up from quantum mechanics (one assumes that there is some local attractive interaction between electrons, and perform a mean field approximation), while the older Ginzburg-Landau theory is derived top-down from thermodynamics (it is assumed that superconductivity can be described by some order parameter, and perform a series expansion of the Helmholtz free energy in the order parameter near the critical point).

REFERENCES

1 W. Meissner and R. Ochsenfeld, Naturwiss. **21**, 787 (1933).
2 F. London and H. London, Proc. Roy. Soc. **A 149**, 71, London (1935).
3 C. Gorter and H.B.G. Casimir, Physica **1**, 306 (1934); C. Gorter, and H.B.G. Casimir, Physik. Z **35**, 963 (1934).
4 R. Meservey and B.B. Schwartz, Superconductivity, R.D. Parks, ed. Dekker, New York, 1969.
5 A.B. Pippard, Proc. R. Soc. **A 216**, 547 (1953).
6 V.L. Ginzburg and L.D. Landau, JETP (URSS) **20**, 1064 (1950).
7 L.D. Landau, Zh. Eksp. Teor. Fiz. **7**, 19 (1937); **7**, 627 (1937); translation into English Collected Papers Of L.D. Landau, edited by D. Ter Haar, Pergamon Press, Oxford, p. 193, 1965 ; also see L.D. Landau and E. Lifchitz, Statistical Physics, Mir Edition, Moscow, Ch. XIV, 1967.
8 P.H. Van Laer and W.H. Keeson, Physica **5**, 993 (1938).
9 B. Bleaney and B.I. Bleaney, Electricity and Magnetism, Oxford University Press, London, 1976.

10 M. Cyrot and D. Pavuna., Introduction to Superconductivity and High-T_c Materials, World Scientific, 1992.

11 P.G. De Gennes, Superconductivity of Metals and Alloys, Benjamin, New York, 1966.

12 J. Bardeen, L.N. Cooper and J.R. Schrieffer, Phys. Rev. **106**, 162 (1957); **108**, 1175 (1957).

13 C. Kittel, Solid State Physics, Bordas, Paris, 1983.

14 B.S. Deaver Jr. and W.M. Fairbank, Phys. Rev. Lett. **7**, 43 (1961).

15 R. Doll and M. Nabauer, Phys. Rev. Letters **7**, 51 (1961).

16 W.A. Little and R.D. Parks, Phys. Rev. Lett. **9**, 9 (1962).

17 J.G. Bednorz and K.A. Müller, Z. Phys. **B 64**, 189 (1986).

18 A.A. Abrikosov, L.P. Gor'kov and I.E. Dzialoshinski, Methods of Quantum Field Theory in Statistical Physics, Dover, New York, 1975.

Chapter 3

Characteristics of High-T_C Superconductors

1. Introduction

The key question about high-temperature superconductivity is the variation of the critical temperature T_C with the material characteristics. In the recent years, several models have been proposed in order to explain the high value of T_C. Even though these models are not unanimously agreed upon within the scientific community, it is not less true that they partly predict the variations of certain parameters. The most realistic theories seem to be those which refine or modify the already existing classical theories (London model, Ginzburg-Landau theory, BCS theory). Of the many surprising facts which have appeared with these new materials, some are[1]: an unusual irreversibility line in the *H-T* plane with a possible fusion of the vortices lattice above this line, an enormous non exponential relaxation time of the magnetic properties and its ageing effects, a disagreement in the estimation of the critical current densities, an almost exponential decrease of the critical current density as a function of the temperature and the field, an anomaly in the profile of the variation of the susceptibility as a function of the field, an unforeseen widening of the resistive transition under field and many others.

These properties, which are the starting point of the present and future interpretations, require a full and detailed knowledge of the structure of these materials. Indeed, although the fundamental interactions which make superconductors of these compounds remain to this day little known, their structural characteristics seem to play a crucial role in the appearance of this phenomenon because almost all of these new materials contain copper oxide planes alternated by insulating or weakly metallic layers. Below the superconducting transition temperature, these planes turn into ideal propagation ways for the charge carriers: the two-dimensionality of the crystalline structure and the mixed valence of the copper appear to be at the origin of the superconducting character of these materials.

Starting from the basic discovery in September 1986 by J.G. Bednorz and K.A. Müller[2] about the $La_{2-x}M_xCuO_{4-y}$ compound with a T_C about 30K, a material previously synthesized at Caen by B. Raveau[3], it was quickly realized that it was possible, on the one hand to obtain critical temperatures of 20 to 40K in the whole family of $La_{2-x}M_xCuO_{4-y}$ with M = Ba, Sr, Ca (see Table 3.1), and on the other hand, the pressure could increase the critical temperature beyond 50K for M = Sr. M.K. Wu et al.[4] and C.W. Chu[5] then simulated a chemical pressure by replacing the lanthanum atoms by yttrium atoms in this family and thus produced in January 1987 the $YBa_2Cu_3O_{7-\delta}$ compound with a T_C about 92K, that is to say with a temperature of transition higher than the nitrogen liquefaction temperature. Shortly after, several compounds with a T_C about 90–94K were discovered in the $RBa_2Cu_3O_x$ family where R is a rare earth except the praseodymium (Pr) and the ytterbium (Yb). In January 1988, H. Maeda et al.[6] and C.W. Chu et al.[7] discovered the $Bi_2Sr_2CaCu_2O_x$ compound with a T_C about 110K and in February of that year, Z.Z. Sheng et al.[8] produced the $Tl_2Ba_2Ca_2Cu_3O_{10}$ compound with a T_C of about 125 K. The whole family of high-T_C compounds of the type $(AO)_mM_2Ca_{n-1}Cu_nO_{2n+1}$ was then discovered where A can be replaced by Tl, Pb, Bi or a mixture of these elements, m = 1 or 2, M being Ba or Sr. Currently, the record of the critical temperatures at ambient pressure is held by the $HgBaCa_{n-1}CunO_{2n+2+\delta}$ compound[9] with a T_C about 133K where n = 3. Under pressure, the critical temperature reaches 150K in the $HgBa_2Ca_2Cu_3O_x$ compound.[10,11]

2. Crystallographic Structures

2.1. $La_{2-x}M_xCuO_{4-y}$ (M = Ba, Sr, Ca) System

At high temperatures, this system crystallizes in a centered tetragonal structure of the type La_2CuO_4 and of the space group I4/mmm[22]. The undoped La_2CuO_4 compound is identical to the cubic perovskite cell $LaCuO_3$ wherein 8 lanthanum atoms occupy the summits of a cube, inside which is located an oxygen atoms octahedron centered on a copper atom. When one combines a lanthanum atom with a divalent atom M (Ba or Sr), the structural transition temperature decreases[22-25]. In the case of the $La_{2-x}Sr_xCuO_{4-y}$ compound, R.M. Fleming et al.[26] showed that for $x > 0.20$, no structural or superconducting transition occurs. The highest superconducting transition temperatures are obtained for $x = 0.15$ where the two transition temperatures tend to approach one the other. The orthorhombic structure admits as cell parameters $a = 5.3420$Å, $b = 5.3547$Å and $c = 13.1832$Å. The separation between two CuO_2 planes is equal to $c/2$ that is to say of the order of 6.65Å[27]. The $La_{2-x}M_xCuO_{4-y}$ system contains a CuO_2 plane per unit-cell. These planes are separated by two $La_{2-x}M_xO$ planes. The substitution of the trivalent lanthanum by

Table 3.1. Critical temperatures at ambient pressure of some high-T_C superconductors.

Formula	T_c(K)	Crystal structure	Unit-cell parameters	Space group	Reference
$(Nd,Ce)_2CuO_{4-\delta}$	24	Tetra	a = 3.95 c = 12.07	I4/mmm	[12]
$(Pb,Cu)(Eu,Ce)_2(Sr,Eu)_2Cu_2O_9$	25	Tetra	a = 3.80 c = 29.60	I4/mmm	[13]
$Pb_2(Sr,La)_2Cu_2O_6$	32	Ortho	a = 5.3119 b = 5.4140 c = 12.629	Pman	[14]
$Bi_2Sr_2(Gd,Ce)_2Cu_2O_{10}$	34	Tetra	a = 3.85 c = 17.88	P4/nmm	[15]
$(La,Sr,Ca)_3Cu_2O_6$	58	Tetra	a = 3.8208 c = 19.5993	I4/mmm	[16]
$(Sr,Ca)_5Cu_4O_{10}$	70	Tetra	a = 3.86 c = 43.0	I4/mmm	[17]
$YBa_2Cu_4O_8$	80	Ortho	a = 3.8411 b = 3.8718 c = 27.240	Ammm	[18]
$Tl_2Ba_2CuO_6$	95	Tetra	a = 3.8637 c = 23.1392	I4/mmm	[19]
$HgBa_2CuO_{4+\delta}$	98	Tetra	a = 3.875 c = 9.5132	P4/mmm	[20]
$TlBa_2CaCu_2O_{7-\delta}$	103	Tetra	a = 3.8566 c = 12.754	P4/mmm	[21]

the divalent cation M introduces charge carriers in the conduction band of CuO_2 planes[28]. In the La_2CuO_4 compound, the formula of neutrality is written: $(La^{3+})_2Cu^{2+}(O^{2-})_4$. The substitution of La by a divalent cation (Ba, Sr, Ca), which leads to values of T_C beyond 40K, has led some researchers to consider the valences Cu^{3+}, O^{1-} et Cu^{1+} [29].

2.2. RBa$_2$Cu$_3$Ox System

The $YBa_2Cu_3O_{6+x}$ compound for which the partial concentration in oxygen x is null is an insulator. Its unit-cell is formed by a stacking according to the c axis of three perovskites cells consisting in a sequence of copper-oxygen planes. By addition of oxygen atoms, this compound becomes metallic for $x > 0.4$ and superconductor below its critical temperature T_C. The ions of oxygen of the Cu-O chains attract the electrons of the CuO_2 plane and increase the number of holes in these planes as well as the critical temperature value. A maximal value of the critical temperature of 94K is reached for $x \sim 0.93$. At $x = 1.0$, the critical temperature falls to 92K. The optimal doping $YBa_2Cu_3O_{6.9}$ compound is commonly called YBCO or merely "123". Its average structure is orthorhombic, of the space group Pmmm and of the parameters $a = 3.8227$Å, $b = 3.8872$Å and $c = 11.6802$Å[30]. Its unit-cell is structured as follows (see Fig. 3.1):

- One plane Cu-O which contains two oxygen lacunas. In this plane, the atom of copper Cu(1) is surrounded by four oxygen ions. This plane forms the CuO chains.
- One plane Ba-O.
- One plane Cu-O where the atom of copper Cu(2) is surrounded of 5 ions of oxygen. This is the famous CuO_2 plane.
- One plane of yttrium that possesses 4 oxygen lacunas.

The rest of the structure is symmetrical in relation to the yttrium ion which can be replaced by one of the rare earths except Pr and Yb without losing its superconducting properties, forming the RBa_2Cu_3Ox system. The copper atom occupies two different sites: Cu(1) in the center of a square CuO_4 and Cu(2) inside a pyramid to square basis CuO_5. The distance between two CuO_2 biplanes is about 11.7Å[27]. The separation made by the yttrium ion gives to this structure a two-dimensional character. The Cu-O chains are formed of the copper atom Cu(1) and the oxygen atom O(1) along the **b**-axis at $(0,0,0)$ and $(0,1/2,0)$, respectively (see Fig. 3.1). The oxygen atom O(4) is located underneath and above Cu(1) at $(0,0,z_{O(4)})$. The distance Cu(1)–O(4) is the smallest distance copper-oxygen of the structure. The copper atom Cu(1) has a number of coordination equal to 4, and every oxygen atom O(1) is surrounded by two copper atoms Cu(1). The CuO_2 planes contain the Cu(2) site at $(0,0,z_{Cu(2)})$ and the O(2) neighboring sites at $(1/2,0,z_{O(2)})$ and O(3) at $(0,1/2,x_{O(3)})$. The different heights, given in c units, are: $z_{Cu(2)} = 0.3574$, $z_{O(2)} = 0.3767$, $z_{O(3)} = 0.3804$ and $z_{O(4)} = 0.1542$[22]. These values, as besides all the structural parameters, depend on the stoichiometry and the history of the sample[31-35].

These planes are therefore undulating, with two oxygen atoms O(2) and O(3) in the plane and the copper atom Cu(2) a little above at 0.022 c = 0.25Å, toward the O(4) site; however, the distance Cu(2)–O(4) is great (2.38Å) in relation to the distance Cu(1)–O(4) (1.78Å). The site Cu(1) being taken at the origin, the site of the yttrium atom is situated at $(1/2,1/2,1/2)$ and the site of the barium atom at $(1/2,1/2,z_{Ba})$ with $z_{Ba} = 0.1895c$.

The formal valence of the $YBa_2Cu_3O_{7-\delta}$ compound is: $Y^{3+}(Ba^{2+})_2(Cu^{2+})_2$ $Cu^{3+}(O^{2-})_7$, the trivalent copper ion is situated in the chains. Because this compound is metallic, the concept of valence is problematic.

2.3. Thallium and Bismuth Compounds

The structure of the thallium compound $Tl_2Ba_2CaCu_2O_8$ (Tl2212) is centered tetragonal, of the space group I4/mmm[36]. The unit-cell contains two CuO_2 planes separated by the calcium ions, and two planes TlO separated of the CuO_2 planes by a BaO plane. Its unit-cell parameters are: $a = 3.8550$Å and

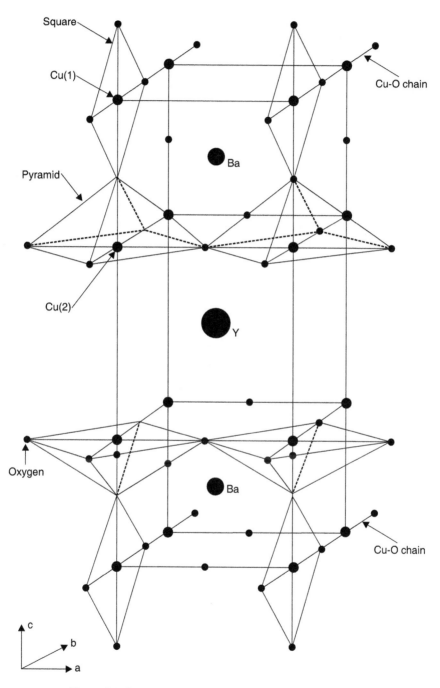

Figure 3.1. Structure of the superconducting oxide YBa$_2$Cu$_3$O$_{7-d}$.

$c = 29\ 318\text{Å}$[37] and the positions of the atoms are summarized in Table (3.2).

The $Bi_2Sr_2CaCu_2O_8$ (Bi2212) compound has an identical structure to that of Tl2212. The unit-cell contains two planes Cu-O(1) and two planes Bi-O(3). The two Cu-O(1) planes are separated by the ions of Ca, whereas these two-planes themselves are separated of the Bi-O(3) two-planes by a plane of Sr-O(2).

These two compounds belong, in fact, to a hierarchy of structures of the types $Bi_2Sr_2Ca_{n-1}Cu_nO_{4+2n}$ and $Tl_2Ba_2Ca_{n-1}Cu_nO_{4+2n}$ with critical temperatures increasing with n until $n = 3$. These structures contain n CuO_2 planes separated by the Ca^{2+} ions[36,39-42]. There is a limiting structure with this series for $n \to \infty$; it consists of a single plane of CuO_2 per unit-cell. These planes are separated along the c-axis by a divalent cation[43]. The fact that the compounds of this series are metallic and especially high-T_C superconductors, it is problematic to establish their neutrality formulas.

Table 3.2. Positions of the various atoms of the compound $Tl_2Ba_2CaCu_2O_8$ (after H. Shaked et al., 1994[38]).

N°	Label	x	y	z
1.	Ca	0.0000	0.0000	0.0000
2.	Tl	0.5000	0.5000	0.2136
3.	Ba	0.0000	0.0000	0.1218
4.	Cu	0.5000	0.5000	0.0540
5.	O(1)	0.0000	0.5000	0.0531
6.	O(2)	0.5000	0.5000	0.1461
7.	O(3)	0.6040	0.5000	0.2815

2.4. Artificial Multilayers

These are heterogeneous structures that are synthesized by starting from the high-T_C superconductors[44-45]. They consist of a stacking of layers S/X ... $/S/X$ ... where S denotes a high-T_C superconductor and X a metallic material, an insulating material or a superconductor with a weaker T_C and a structure close to that of S. Let us mention for example the super lattices: $YBa_2Cu_3O_7/DyBa_2Cu_3O_7$[46] and $YBa_2Cu_3O_7/PrBa_2Cu_3O_7$[47-49]. These structures permit the study of the variations of the characteristic parameters of the high-T_C superconductors as a function of the composition and the thickness of the layers. Two other systems are also concerned with the study of the anisotropy of the high-T_C superconductors: the multilayers of the conventional superconductors[50] and the organic superconductors[51-54] (in particular those of the family BEDT-TTF[55]) which have a layered structure similar to that of the high-T_C superconductors.

3. Properties

With the exception of some compounds such as $BaPb_{1-x}Bi_xO_3$ ($T_C \sim 13K$, $x = 0.25$)[56-58], $Ba_{1-x}K_xBiO_3$ ($T_C > 30K$)[59-61] and $LiTi_2O_4$ ($T = 12K$)[62-63] which do not contain copper atoms, all the other compounds are metallic oxides which have stratified structures composed of CuO_2 planes separated by insulating or weakly metallic layers. This layered structure induces in these materials a strong anisotropy of all their superconducting properties. The electrical conductivity is very high along the *ab* planes whereas it is much weaker in the direction perpendicular to them. The critical current is very high when the current circulates along the layers and it is much weaker in the perpendicular direction. The critical fields are higher in the direction of the layers than in the perpendicular direction. In addition to high transition temperatures, the cuprates are characterised by extraordinarily high upper critical fields B_{C2} which can even exceed 100T at zero temperature. However, the useful magnetic field range of these superconductors is limited by the irreversibility line well below the upper critical field. Above this irreversibility line, the pinning force for the vortices is negligible, and hence the material is a superconductor with vanishing critical current density. Whereas the most oxides are insulators, the high-T_C superconductors are metallic oxides (in fact, they gradually become so by doping). The high-T_C superconductors are ceramics which differ from coffee or tea services only by the color and by the fact that below T_C, the first mentioned become superconductors. The weak pinning in these materials is closely connected to the extremely short coherence length of less than 1nm along the *c* direction.

For a detailed discussion of these aspects and an extensive review of the high-T_C superconductors' characteristics, one can consult the works of Y. Iye[64]; and D.R. Harshman and A.P. Mills[65].

4. Elaboration

4.1. Introduction

The manufacture of the ceramic superconducting materials in the form of thin layers, ribbons, wires or massive products requires the control of the elaboration and the implementation of the basic products. These basic materials can be gaseous products (for example organometallic mixtures of compounds in vapor phase), liquids or gels (molten mixtures, aqueous solutions, barbotine, etc.) or solid phases, notably powders composed of precursory chemical products or those that have already reacted in the form of pure or mixed ceramic oxides[66]. In this section, we give a brief description of some elaboration techniques of the most used high-T_C superconductors.

4.2. Thin Films

The thin films of the high-T_C superconducting compounds are prepared in a stage which simultaneously combines the formation by deposition on substrate and the crystallization of the desired phase. The main parameters to be controlled in order to obtain films of good quality are the crystallinity, the orientation of grains and the stoichiometry. The choice of the substrate is very important because, on the one hand the diffusion phenomena at the level of the film-substrate interface can involve a loss of the superconductivity, and on the other hand the crystalline structure of the substrate (cell parameter and orientation) facilitates the obtaining of layers of good quality. Thus, the best films were obtained by deposition of $SrTiO_3$, ZrO_2, MgO, $LaAlO_3$, $LaGaO_3$ and $KNbO_3$ on the single crystals. In the area of the single crystalline thin layers of $YBaCu_3O_{7-\delta}$, critical current densities of the order of $10^7A/cm^2$ at 4.2K, and of 3 to $5 \times 10^6A/cm^2$ at 77K were obtained. The dependence of the magnetic field is very weak even at 77K, the temperature decrease is normal since J_C must vanish at $T = T_C$. In the thin films of BiSrCaCuO, critical current densities of $2 \times 10^7A/cm^2$ were obtained below 40K while J_C does not exceed $2 \times 10^6A/cm^2$ at 77K. The application of a magnetic field of 0.5 Tesla at 77K removes the critical current. This weakening of J_C under magnetic field seems to be intrinsic to the BiSrCaCuO and TlBaCaCuO compounds. The main techniques which lead to good results can be summarized as follows:

A. Cathode Sputtering

This technique, extensively used in microelectronics, is based on the following principle: the metal source, called "the target", is placed on an electrode (cathode) and introduced into a vacuum chamber. The substrate on which the thin film must be deposited is placed against the anode, parallel to the cathode, nearly 10cm of this one. A difference of potential between the two electrodes is applied under vacuum. The residual gas (argon) is ionized and the positive ions are attracted by the cathode. Under the impact, the atoms of the target are torn off and come to deposit on the substrate by progressively forming the thin film. Several alternatives of this process exist: continuous diode, radio frequency diode, triode, and magnetron. In the last case, a magnetic field is held parallel to and close to the target in order to facilitate the wrenching of the atoms. It results in a higher deposition rate. The obtained films are of excellent quality.

The cathode sputtering was originally used to obtain layers of niobium or niobium nitride. For a high-T_C superconductor, of complex composition, one uses several targets. For example, Y_2O_3, $BaCO_3$, CuO are used in order to elaborate a sample of YBaCuO. One can also use only one target of YBaCuO with controlled oxidation in situ. An annealing after deposition at a temperature above 800°C optimizes the superconducting properties of the film. Further work is needed to determine the best annealing parameters.

B. Vapor Deposition UHV

This is a multi-source technology i.e. there are as many sources as metallic elements to deposit. The substrate is placed in a vacuum chamber with the sources. The metallic elements are preferably vaporized by electronic bombardment. One recuperates this vapor by condensation on the substrate which is maintained at a temperature between 500°C and 700°C during the deposition. The oxygen is incorporated in the film during the deposition; this source of oxygen can be an outgoing controlled flux of a tube in the vicinity of the substrate or oxygen plasma excited by radiofrequency. The pressure in the chamber is maintained at a value of the order of 10^{-7} Torr during the growth.

C. Pulsed Laser Deposition

The ablation of a source of the composite material (single target) is produced by means of high-energy laser impulses. This technique is similar to the previous one in its principle. It allows realizing very thin films of excellent crystallinity.

D. Chemical Vapor Deposition (CVD)

This is a multisource growth technique in-situ. A chemical reaction is provoked between the different gaseous elements in order to attain the desired deposit. Either a heterogeneous chemical reaction catalyzed by the substrate, or a homogeneous reaction within the gas is obtained, followed by the deposition on the substrate and the crystallization of the compound. However, while the results may be very good for research, none of these techniques are entirely satisfactory for the industrial developments.

For the basic research, single crystalline substrates such as $SrTiO_3$, ZrO_2, MgO are entirely satisfactory. However, they are expensive and marketed as crystals of small size, of the order of $1cm^2$.

For industrial applications, we must realize inexpensive substrates such as silicon wafers in order to solve the problem of the superconducting film-substrate interaction. One then turns to the production of multilayers where a buffer layer, MgO (see Fig. 3.2) allows, firstly, chemically isolating the silicon and the superconductor, and secondly inducing a preferred orientation in the superconducting layer. Furthermore, the magnetron sputtering can impose itself in the short run for the industrial needs, because it is the only method that is economically optimized.

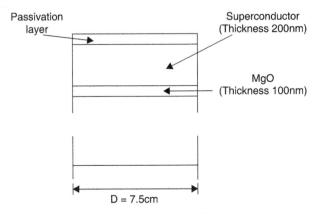

Passivation layer

Superconductor (Thickness 200nm)

MgO (Thickness 100nm)

D = 7.5cm

Figure 3.2. Multilayer film.

4.3. Single Crystals

The most studies on single crystals of the $RBa_2Cu_3O_{7-\delta}$ family reported in the literature were performed on crystals prepared by flux methods[67-70]. In most cases, these crystals exhibited large superconducting transitions and critical temperatures lower than 85K. This behavior is now known. It results from a strong contamination of crystals by the crucible (usually in aluminum) leading to a substitution of the copper atom by Al^{3+} which provokes a strong reduction of the critical temperature T_c[71]. The single crystals of the type $RBa_2Cu_3O_{7-\delta}$ (R = Sm-Tm) can be obtained by the melt grow method as follows: a mixture of oxides of rare earths, of barium and copper carbonate is compressed into pellets which are then deposited in an alumina crucible on a layer of $RBa_2Cu_3O_{7-\delta}$ in order to avoid any contamination with the bottom of the crucible (or in a ceramic crucible). These pellets are then heated for 10 to 20 hours at temperatures slightly below (~ 1 to 2°C) the peritectic transition temperature, determined in advance by quenching experiments (~ 1040°C, 980°C, 970°C for Gd, Y and Er, respectively). Next, the pellets undergo in a first time, a controlled cooling until 900°C at a rate of 5°C/hour and then a second cooling without particular control until room temperature. Single crystals of high quality which can reach 50mg can be easily obtained by a simple cleavage according to the ab planes. Neutron diffraction experiments carried out on samples obtained by this method showed that they were well single crystals[72]; however, due to their growth temperature close to the peritectic decomposition temperature, inclusions of green phases cannot be avoided in the larger crystals. All these crystals are in the final step exposed to a flux of oxygen during 24 to 48 hours at 450°C.

Moreover, single crystals of the type $Bi_2Sr_2CaCu_2O_8$ can be manufactured from a mixture of $Bi_2O_3 + 2\,SrCO_3 + CaCO_3 + 2\,CuO$ heated in a platinum crucible at 900°C for 2 hours. This mixture is cooled as the first step until 800°C at a controlled rate of 5°C/hour and then let cool to room temperature without any particular control. Thin lamellar crystals are obtained, identified by the value of the parameter c deduced by the x-ray diffraction starting from the reflections (001) obtained in a scanning mode θ-2θ using a two circles goniometer. The value of the parameter c which is really measured ($c = 30.86$Å) characterizes the Bi2212 phase[73]. Finally, single crystals of $TlBa_2Ca_2Cu_3O_9$ and $Tl_2Ba_2Ca_2Cu_3O_{10}$ are prepared from the stoichiometric mixtures of Tl_2O_3, BaO_2 or BaO, CaO and CuO, heated in a quartz tube at T ~ 950°C for 3 hours and then slowly cooled to the room temperature at a rate of 5°/hour. The starting oxides are premixed in a dry box and then pressed into pellets and placed in a ceramic crucible. Crystal plates of maximum size, $1 \times 1 \times 0.1mm^3$, are mechanically separated from the obtained mixture and individually characterized by the x-ray diffraction. The most of these plates show a syntactic intergrowth (observed also by S.S.P. Parkin et al.[74]); the single crystals, products of this growth are selected with the following parameters: $c = 15.87$Å for Tl1223 and $c = 35.605$Å for Tl2223 in agreement with the literature[75-76]. The values of the current critical densities of these materials, deducted from the hysteresis cycles for a parallel field to the c-axis at $T = 4.2$K, are about $10^8A/m^2$ for TBCCO, $10^9A/m^2$ for BSCCO and $10^{10}A/m^2$ for YBCO[71].

4.4. Textured Polycrystals

The inter-granular current density is generally very high in single crystals at low temperatures. Unfortunately the dimensions of these materials are very small, of the order of a few millimeters in the best cases[77]. For large scale applications of high-T_c superconductors, it is necessary to have long wires and ribbons with a sufficiently high current density at high fields and temperatures; $J_c (T = 77$K, $B = 3$T) $\geq 10^4A/cm^2$ is a completely achievable challenge.

At present, it seems that among the known high-T_c oxides, only Y(123) has the desired value of $J_c(T,B)$. Unfortunately, the mechanical properties of this material and the conditions of the elaboration of the wires and ribbons make its practical use very difficult. Furthermore, the bismuth oxides (BSCCO, BPSCCO, in particular the (2223) and (2212) phases) exhibit significant deformations during their cooling but a significant critical current in zero fields at 4.2K. However, this current suddenly decreases as a function of the applied field at 77K and becomes almost unusable at $B > 0.2$T. Therefore, it seems interesting and realistic, in the future, to develop bismuth oxides for the applications at 4.2K by hoping for more favorable conditions to solve the problems at 77K or to discover other high-T_c materials which could meet all the requirements. It is to be noted that even at 4.2K further work is still needed although very encouraging results on the bismuth materials were established by a number of researchers throughout the world. Among the most used techniques to manufacture these textured materials are the melt grow method, the directional solidification technique, texturing by induced magnetic field and the technique of the tube powder.

REFERENCES

1   A.P. Malozemoff, Physical Properties of High Temperature Superconductors, ed. D. Ginsberg, World Scientific, Singapore, 1989.
2   J.G. Bednorz and K.A. Müller, Z. Phys. **B 64**, 189 (1986).
3   B. Raveau, La Vie des Sciences **4**, 93 (1987).
4   M.K. Wu, J.R. Ashburn, C.J. Torng, P.H. Hor, R.L. Meng, L. Gao, Z.J. Huang, Y.G. Wang and C.W. Chu, Phys. Rev. Lett. **58**, 907 (1987).
5   C.W. Chu, P.H. Hor, R.L. Meng, L. Gao, Z.L. Huang and T.Q. Wang, Phys. Rev. Lett. **58**, 405 (1987).
6   H. Maeda, Y. Tanaka, M. Fukutomi and T. Asano, Jpn. J. Appl. Phys. Lett. **27**, L209 (1988).
7   C.W. Chu, J. Bechtold, L. Gao, P.H. Hor, Z.J. Huang, R.L. Meng, Y.Y. Sun, Y.Q. Wang and Y.Y. Xue, Phys. Rev. Lett. **60**, 941 (1988).
8   Z.Z. Sheng and A.M. Hermann, Nature **332**, 55 (1988).
9   S.N. Putilin, E.V. Antipov, O. Chmaissem and M. Marezio, Nature, **362**, 226 (1993); S.N. Putilin, E.V. Antipov and M. Marezio, Physica **C 212**, 266 (1993).
10 C.W. Chu, L. Gao, F. Chen, Z.J. Huang, R.L. Meng and Y.Y. Xue, Nature **365**, 323 (1993).
11 M. Numez-Rugueiro, J.L. Tholence, E.V. Antipov, J.J. Capponi and M. Marezio, Science **262**, 97 (1993).
12 J.D. Jorgensen, M.A. Beno, D.G. Hinks, L. Soderholm, K.J. Volin, R.L. Hitterman, J.D. Grace, I.K. Schuller, C.U. Segre, K. Zhang and M.S. Kleefisch, Phys. Rev. **B 36**, 3608 (1987).
13 Y. Yokura, H. Takagi and S. Uchida, Nature **337**, 345 (1989).
14 T. Maeda, K. Sakuyama, S. Koriyama, A. Ichinose, H. Yamouchi and S. Tanaka, Physica **C 169**, 133 (1990).
15 F.J.M. Benschop, W.T. Fu and W.J.A. Maaskant, Physica C **184**, 311 (1991).
16 T. Arima, Y. Tokura, H. Takagi, S. Ushida, R. Beyers and J.B. Torrance, Physica C **168**, 79 (1990).
17 R.J. Cava, A. Santoro, J.J. Krajewski, R.M. Flemming, J.V. Wasczak, W.F. Peck and P. March, Physica C **172**, 138 (1990).
18 P. Fischer, J. Karpinski, E. Kaldis, E. Jilek and S. Rusieck, Solid State Comm. **69**, 531 (1989).
19 F. Izumi, E. Takayama-Muromachi, A. Fujimori, T. Kamiyama, H. Asano, J. Akimitsu and H. Sawa, Physica C **158**, 440 (1989).
20 J.L. Wagner, P.G. Radaelli, D.G. Hinks, J.D. Jorgensen, J.F. MiTChell, B. Dabrowski, G.S. Knapp and M.A. Beno, Physica C **210**, 311 (1989).
21 B. Morosin, D.S. Ginly, P.F. Hlava, M.J. Carr, R.J. Baughman, J.E. Schirber, E.L. Venturini and J.F. Kwak, Physica C **152**, 413 (1988).
22 J.D. Jorgensen, M.A. Beno, D.G. Hinks, L. Soderholm, K.J. Volin, R.L. Hitterman, J.D. Grace, I.K. Schuller, C.U. Segre, K. Zhang and M.S. Kleefisch, Phys. Rev. **B 36**, 3608 (1987).
23 R.J. Cava, A. Santoro, D.W. Johnson, Jr. and W.W. Rhodes, Phys. Rev. **B 35**, 6716 (1987).
24 S.C. Moss, K. Forster, J.D. Axe, H. You, D. Hohlwein, D.E. Cox, P.H. Hor, R.L. Meng and C.W. Chu, Phys. Rev. **B 35**, 7195 (1987).

25   D. Mck. Paul, G. Balakrishnan, N.R. Bernhoeft, W.I.F. David and W.T.A. Harrison, Phys. Rev. Lett. **58**, 1976 (1987).
26   R.M. Fleming, B. Batlogg, R.J. Cava and E.A. Rietman, Phys. Rev **B 35**, 7191 (1987).
27   P. Pugnat, Thesis, Joseph Fourier University, Grenoble I, France, 1995.
28   A.M. Ettouhami, Thesis, Joseph Fourier University, Grenoble I, France, 1994.
29   W.E. Pickett, Rev. Mod. Phys. **61**, 433 (1989).
30   J.D. Jorgensen, B.W. Veal, A.P. Paulikas, L.J. Nowicki, G.W. Crabtree, H. Clauss and W.K. Kwok, Phys. Rev. **B 41**, 1863 (1990).
31   J.J. Capponi, C. Chaillout, A.W. Hewat, P. Lejay, M. Marezio, N. Nguyen, B. Raveau, J.L. Soubeyroux, J.L. Tholence and R. Tournier, Europhys. Lett. **3**, 1301 (1987).
32   M. François, E. Walker, J.L. Jorda, K. Yvon and P. Fischer, Solid State Commun. **63**, 1149 (1987).
33   Y. Le Page, W.R. McKinnon, J.M. Tarascon, L.H. Greene, G.W. Hull and D.M. Hwang, Phys. Rev. **B 35**, 7245 (1987).
34   Q.W. Yan, P.L. Zhang, L .Jin, Z.G. Shen, J.K. Zhao, Y. Ren, Y.N. Wei, T.D. Mao, C.X. Liu, T.S. Ning, K. Sun and Q.S. Yang, Phys. Rev. **B 36**, 5599 (1987).
35   A. Williams, G.H. Kwei, R.B. Von Dreele, A.C. Larson, I.D. Raistrick and D.L. Bish, Phys. Rev. **B 37**, 7960 (1988).
36   M.A. Subramanian, C.C. Torardi, J. Gopalakrishnan, P.L. Gai, J.C. Calabrese, T.R. Asker, R.B. Flippen and A.W. Sleight, Science **242**, 249 (1988).
37   M.A. Subramanian, J.C. Calabrese, C.C. Torardi, J. Gopalakrishnan, T.R. Askew, R.B. Flippen, K.J. Morrisev, U. Chowdhry and A.W. Slight, Nature **332**, 420 (1988).
38   H. Shaked, P.M. Keane, J.C. Rodriguez, F.F. Owen, R.L. Hitterman and J.D. Jorgensen, Crystal Structures of High-T$_c$ Superconducting Copper-Oxides, Elsevier Sciences, Amsterdam, Netherlands, 1994.
39   P. Haldar, K. Chen, B. Maheswaran, A. Roig-Janicki, N.K. Jaggi, R.S. Markiewicz and B.C. Griessen, Science **241**, 1198 (1988).
40   S.S.P. Parkin, V.Y. Lee, A.I. Nazzal, R. Savoy, R. Beyers and S.J. La Placa, Phys. Rev. Lett. **61**, 750 (1988).
41   C.C. Torardi, M.A. Subramanian, J.C. Calabrese, J. Gopalakrishnan, K.J. Morrissey, T.R. Askew, R.B. Flippen, U. Chowdhry and A.W. Sleight, Science **240**, 631 (1988).
42   J.B. Torrance, Y. Tokura, S.J. La Placa, T.C. Huang, R.J. Savoy and A.I. Nazzal, Solid State Commun. **66**, 703 (1988).
43   T. Siegrist, S.M. Zahurak, D.W. Murphy and R.S. Roth, Nature (London) **334**, 231 (1988).
44   A. Gupta, R. Gross, E. Olsson, A. Segmüller, G. Koren and C.C. Tsuei, Phys. Rev. Lett. **64**, 3191 (1990).
45   Q. Li, C. Kwon, X.X. Xi, S. Bhattacharva, A. Walkenhorst, T. Venkatesan, S.J. Hagen, W. Jiang and R.L. Greene, Phys. Rev. Lett. **69**, 2713 (1992).
46   J.M. Triscone, M.G. Karkut, L. Antognazza, O. Brunner and O. Fisher, Phys. Rev. Lett. **63**, 1016 (1989).
47   U. Poppe, P. Prieto, F. Schubert, H. Saltuer and K. Urban, Solid State Commun. **71**, 569 (1989).
48   Q. Li, X.X. Xi, X.D. Wu, A. Inam, S. Vadlamannati and W.L. McLean, Phys. Rev. Lett. **64**, 3086 (1990).
49   D.H. Lowndes, D.P. Norton and J.D. Budai, Phys. Rev. Lett. **65**, 1160 (1990).
50   L. Chang and B. Griessen ed., Synthetic Modulated Structures, Academic, New York, 1984.
51   T. Ishiguro and K. Yamaji, Organic Superconductors, Series in Solid State Sciences, Vol. 88, Springer, Berlin, 1990.
52   D. Jérome, Science **252**, 1509 (1991).
53   A.E. Underhill, J. Mater. Chem. **2**, 1 (1992).
54   L.N. Bulaevskii, Adv. Phys. **37**, 443 (1988).
55   P.A. Mansky, P.M. Chaikin and R.C. Haddon, Phys. Rev. Lett. **70**, 1323 (1993).
56   A.W. Sleight, J.L. Gillson and P.E. Bierstedt, Solid St. Commun. **17**, 27 (1975).

57 L.F. Mattheiss and D.R. Hamann, Phys. Rev. **B 26**, 2686 (1982).

58 L.F. Mattheiss and D.R. Hamann, Phys. Rev. **B 28**, 4227 (1983).

59 L.F. Mattheiss, E.M. Gyorgy and D.W. Johnson, Jr., Phys. Rev. **B 37**, 3745 (1988).

60 R.J. Cava, B. Batlogg, J.J. Krajewski, R. Farrow, L.W. Rupp, A.E. White, K. Short, W.F. Pexk and T. Kometani, Nature (London) **332**, 814 (1988).

61 L.F. Mattheiss and D.R. Hamann, Phys. Rev. Lett. **60**, 2681 (1988).

62 S. Satpathy and R.M. Martin, Phys. Rev. **B 36**, 7269 (1987).

63 S. Massidda, J. Yu and A.J. Freeman, Phys. Rev. **B 38**, 11352 (1988).

64 Y. Iye, Comments Cond. Mat. Phys. **16**, 89 (1992).

65 D.R. Harshman and A.P. Mills, Phys. Rev. **B 45**, 10684 (1992).

66 G. Bronca (coordinator of the group), Applications of Superconductivity, Synthesis Report of the group "Superconductivity" French Observatory of advanced techniques, Masson, Paris, 1990.

67 L.F. Schneemeyer, J.V. Waszczak, T. Siegrist, R.B. Van Dover, L.W. Rupp, B. Batlogg, R.J. Cava and D.W. Murphy, Nature **328**, 601 (1987).

68 H.J. Scheel and J. Less Com. Met. **151**, 199 (1989).

69 M. Maeda, M. Kadoi and T. Ikeda, Jpn. J. Appl. Phys. **28**, 1417 (1989).

70 T. Wolf, W. Goldacker, B. Obst, G. Roth and R. Flukiger, J. Cryst. Growth **96**, 1010 (1989).

71 J.L. Tholence, M. Saint-Paul, O. Laborde, P. Monceau, M. Guillot, H. Noël, J.C. Levet, M. Potel, J. Padiou and P. Gougeon, Studies of High Temperature Superconductors, Vol. 6, ed. A. Narlikar, Nova Science Publishers, New York, 1990.

72 P. Burlet, C. Vettier, M. Jurgens, J.Y. Henry, J. Rossat-Mignod, H. Noël, M. Potel, P. Gougeon and J.C. Levet, Physica **C 153**, 1115 (1988).

73 M. Hervieu, C. Michel, C. Domenges, Y. Laligant, A. Lebail, G. Forey and B. Raveau, Mod. Phys. Lett. **B 2**, 491 (1988).

74 S.S.P. Parkin, V.Y. Lee, E.M. Engler, A.I. Nazzal, T.C. Huang, G. Gorman, R. Savoy and R. Beyers, Phys. Rev. Lett. **60**, 2539 (1988).

75 B. Morosin, D.S. Ginley, P.F. Hlava, M.J. Carr, R.J. Baughman, J.E. Schirber, E.L. Venturini and J.F. Kwak, Physica **C 152**, 413 (1988).

76 B. Morosin, D.S. Ginley, J.E. Schirber and E.L. Venturini, Physica **C 156**, 587 (1988).

77 S. Senoussi, J. Phys. III (France) **2**, 1041 (1992).

Chapter 4

Phenomenoligical Theories of the Anisotropic Superconductors

1. Anisotropic Ginzburg-Landau Model

1.1. Free Energy

As the coherence length $\xi^c(T)$, perpendicular to the ab planes of the high-T_C superconductors, is very low at low temperature, these materials can be described by a continuum model only at a temperature range defined by $\xi^c(T) \gg d$ (d is the distance between the planes). In this temperature range near T_C, these materials with the laminated structures can be adequately described by the anisotropic Ginzburg-Landau model. To go from the isotropic model to this model, it is sufficient to replace the electron mass m in the expression of the energy by a tensor of effective masses $(m)_{ij}$ describing the electronic anisotropy of the material[1]. In a superconductor with a uniaxial symmetry, supposed case of the high-T_C materials, the orthogonal system of $Oxyz$ axes is chosen such that the **Oz** axis is parallel to the c-axis of the crystal. The tensor of the masses is then reduced to a diagonal tensor whose principal values are: $m^a = m^b = m^{ab}$, effective masses of the Cooper pairs in the superconductor's planes, and m^c effective mass in the direction perpendicular to the planes:

$$\overline{\overline{\sigma}} = \begin{vmatrix} m^{ab} & 0 & 0 \\ 0 & m^{ab} & 0 \\ 0 & 0 & m^c \end{vmatrix} \tag{4.1}$$

The free energy is written:

$$F_S(T) = \int_{(V)} d^3\mathbf{r} \left[\alpha(T)|\psi|^2 + \frac{\beta}{2}|\psi|^4 + \right.$$

$$\frac{\hbar^2}{2m^{ab}} \left| \left(\nabla_{//} - \frac{2i\pi}{\Phi_0} \mathbf{A}_{//} \right) \psi \right| \tag{4.2}$$

$$\left. + \frac{\hbar^2}{2m^c} \left| \left(\frac{\partial}{\partial z} - \frac{2i\pi}{\phi_0} A_z \right) \psi \right|^2 \right] + \int d^3\mathbf{r} \frac{\hbar^2}{8\pi}$$

where $\mathbf{A}_{//} = (A_x, A_y)$, $\nabla_{//} = (\partial/\partial x, \partial/\partial y)$ and $\mathbf{h} = \nabla \times \mathbf{A}$. The minimization of this energy leads to the anisotropic Ginzburg-Landau equations:

$$\frac{1}{2m^c} \left(-i\hbar\nabla_z + \frac{eA_z}{c} \right)^2 \psi +$$

$$\frac{1}{2m^{ab}} \left(-i\hbar\nabla_{//} + \frac{e\mathbf{A}_{//}}{c} \right)^2 \psi + \tag{4.3}$$

$$\alpha\psi + \beta|\psi|^2\psi = 0$$

$$\mathbf{j}_{//} = -\frac{e\hbar}{2im^{ab}} (\psi^* \nabla_{//}\psi - \psi\nabla_{//}\psi^*) -$$

$$\frac{e^2}{mc}|\psi|^2 \mathbf{A}_{//} \tag{4.4a}$$

$$j_\perp = -\frac{e\hbar}{2im^c} (\psi^* \partial_z\psi - \psi \partial_z\psi^*) - \frac{e^2}{m^c c}|\psi|^2 A_z \tag{4.4b}$$

In the absence of field, one obtains the value of the order parameter:

$$\psi_0 = \sqrt{\frac{|\alpha(T)|}{\beta}} \tag{4.5}$$

where $\alpha(T) = \alpha_0 (T - T_c)$, and the expression of the condensation energy:

$$\frac{H_C^2(T)}{8\pi} = \frac{\alpha^2(T)}{2\beta} \tag{4.6}$$

which are both similar to the isotropic case (see Eq. 4.39 and Eq. 4.41).

1.2. Coherence Lengths

This form of the free energy introduces two coherence lengths, a length parallel to the *ab* planes:

$$\xi^{ab}(T) = \frac{\hbar}{\sqrt{2\,m^{ab}\,|\alpha(T)|}} \tag{4.7}$$

and a second perpendicular to these:

$$\xi^{c}(T) = \frac{\hbar}{\sqrt{2\,m^{c}\,|\alpha(T)|}} \tag{4.8}$$

whose ratio defines the anisotropy parameter γ:

$$\gamma = \frac{\xi^{ab}}{\xi^{c}} = \sqrt{\frac{m^{c}}{m^{ab}}} \geq 1 \tag{4.9}$$

1.3. Penetration Depths

The free energy within the limit of London, for which the applied field H is such as $H_{C1} \ll H \ll H_{C2}$, is written:

$$F_{L} = \frac{1}{8\pi} \int_{(V)} d^{3}\mathbf{r}\ [\mathbf{h}^{2} +$$
$$\lambda_{ab}^{2}\,(\operatorname{rot}\mathbf{h})_{//}^{2} + \lambda_{c}^{2}\,(\operatorname{rot}\mathbf{h})_{\perp}^{2}\,] \tag{4.10}$$

where λ^{ab} and λ^{c} correspond to the respective penetration depths for the directions parallel and perpendicular to the *ab* planes.

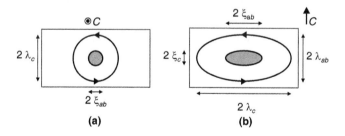

Figure 4.1. Anisotropy of the coherence length and the penetration depth in a superconductor with a uniaxial symmetry. (a) The system being isotropic parallel to the layers, a vortex along the *c*-axis will be isotropic (characteristic lengths ξ_{ab} and λ_{ab}). (b) On the other side, a parallel vortex to the layers will be anisotropic, having the shape of an ellipse of which the ratio of the axes is proportional to γ. More generally, the vortices have an anisotropic structure when the magnetic field makes a nonzero angle θ with the *c*-axis of the crystal.

The penetration depths λ^{ab} and λ^{c} are defined by:

$$\lambda^{ab} = \sqrt{\frac{m^{ab}}{4\pi\,e^{2}\,\psi_{0}^{2}}} \tag{4.11}$$

$$\lambda^c = \sqrt{\frac{m^c}{4\pi \, e^2 \, \psi_0^2}} \tag{4.12}$$

whose ratio gives:

$$\lambda^c = \gamma \, \lambda^{ab} \tag{4.13}$$

As $\gamma \gg 1$, Equation 4.13 shows that $\lambda^c \gg \lambda^{ab}$. When the vortex is aligned along the *ab* planes, the anisotropy changes its section which takes the form of an ellipse whose axial ratio is proportional to γ. For an aligned vortex along the c-axis, its section remains circular exactly as in the case of an isotropic vortex, of characteristic lengths ξ^{ab} and λ^{ab} (see Fig. 4.1).

1.4. Ginzburg-Landau Parameters

The Ginzburg-Landau parameters, κ^{ab} and κ^c which correspond to a field parallel and perpendicular to the ab planes are defined by:

$$\kappa^{ab} = \sqrt{\frac{\lambda^{ab} \, \lambda^c}{\xi^{ab} \, \xi^c}} \qquad \text{and} \qquad \kappa^c = \frac{\lambda^{ab}}{\xi^{ab}} \tag{4.14}$$

1.5. Critical Fields

The anisotropy of the superconducting properties is also reflected in the expressions of the critical fields H_{C1} and H_{C2} which will depend this time on the orientation of the applied field. The angular dependence of the lower critical field $H_{C1}(\theta)$, where θ is the angle between the applied magnetic field **H** and the **c**-axis of the crystal, can be described within the limit of London, by neglecting the contribution of the vortex core, by[2,3]:

$$H_{C1}(\theta) = \frac{\Phi_0}{4\pi \, (\lambda^{ab})^2} \, \frac{\ln\left[\dfrac{\kappa^{ab}}{\varepsilon(\theta)}\right]}{\sqrt{\cos^2\theta + \gamma^2 \sin^2\theta}} \tag{4.15}$$

where:

$$\varepsilon(\theta) = \sqrt{\sin^2\theta + \gamma^2 \cos^2\theta} \tag{4.16}$$

whose limiting values are:

$$H_{C1}^c = \frac{\Phi_0}{4\pi \, (\lambda^{ab})^2} \, \ln\left(\frac{\lambda^{ab}}{\xi^{ab}}\right) \quad \text{for } \mathbf{H}//\mathbf{c} \tag{4.17}$$

and:

$$H_{C1}^{ab} = \frac{\Phi_0}{4\pi\lambda^{ab}\lambda^c} \ln\left(\frac{\lambda^{ab}\lambda^c}{\xi^{ab}\xi^c}\right)^{1/2} \quad \text{for } \mathbf{H}//(\mathbf{a,b}) \tag{4.18}$$

such as:

$$H_{C1}^{ab} \ll H_{C1}^c \tag{4.19}$$

The thermodynamic field H_C preserves the same expression as the isotropic case, namely:

$$H_c(\theta) = H_C \tag{4.20}$$

The upper critical field $H_{C2}(\theta)$ is given by:

$$H_{C2}(\theta) = \frac{\Phi_0}{2\pi\,\xi_{ab}^2\sqrt{\cos^2\theta + \gamma^{-2}\sin^2\theta}} \tag{4.21}$$

whose limiting values are:

$$H_{C2}^c = \frac{\Phi_0}{2\pi\left(\xi^{ab}\right)^2} \quad \text{for } \mathbf{H}//\mathbf{c} \tag{4.22}$$

and:

$$H_{C2}^{ab} = \frac{\Phi_0}{2\pi\,\xi^{ab}\xi^c} \quad \text{for } \mathbf{H}//(\mathbf{a,b}) \tag{4.23}$$

whose ratio is:

$$\frac{H_{C2}^c}{H_{C2}^{ab}} = \frac{\xi^c}{\xi^{ab}} \leq 1 \tag{4.24}$$

This shows that the critical field parallel to the ab planes is higher than that which is perpendicular to them. The critical fields $H_{Ci}(\theta)$, where $i = 1, 2$, can be written, to logarithmic accuracy with regard to $H_{C1}(\theta)$, as follows:

$$\left[\frac{H_{Ci}(\theta)\cos\theta}{H_{Ci}^c}\right]^2 + \left[\frac{H_{Ci}(\theta)\sin\theta}{H_{Ci}^{ab}}\right]^2 = 1 \tag{4.25}$$

where $H_{Ci}^c = H_{Ci}(\theta = 0)$ and $H_{Ci}^{ab} = H_{Ci}(\theta = \pi/2)$. Equation 4.25 shows that the curves $H_{Ci}(\theta)$ in the plane ($H_X = H\sin\theta$, $H_Z = H\cos\theta$) form ellipses (see Fig. 4.2). The angular dependence of the third upper critical field $H_{C3}(\theta)$, which characterizes the destruction of the surface superconductivity, is given by:

$$H_{C3}(\theta) \sim 1.69 \times \frac{\sqrt{2}\,\kappa^{ab}\,H_C}{\varepsilon(\theta)} \qquad (4.26)$$

The critical field H_{C3} represents in this case the geometric average $[(H_{C3}^{ab})^2\,H_{C3}^c]^{1/3}$.

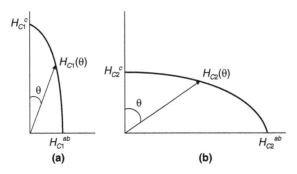

Figuer 4.2. The critical fields form in the plane ($H\,sin\theta$, $H\,cos\theta$) ellipses. (a) Lower critical field $Hc_1(\theta)$. (b) Upper critical field $Hc_2(\theta)$. The two figures are not at the same scale.

1.5.1. Example of the Experimental Determination of H_{C1}

The lower critical field H_{C1}^c, of a single crystal of $YBa_2Cu_3O_{7-\delta}$ for example, can be estimated from the first penetration of the magnetic flux at the boundary between the Meissner state and the mixed state at a field $H_i(T)$ by applying a field parallel to the **c**-axis of the crystal. Indeed, in the emu-cgs system, the induction B^c (parallel to the **c**-axis) inside the sample is written:

$$B^c = [H + 4\pi\,(1 - N^c)\,M^c] = H_i^c + 4\pi M^c \qquad (4.27)$$

Knowing that in the Meissner phase one has $B^c = 0$, it becomes:

$$H_{C1}^c(T) \sim H_i^c(T) = \frac{H}{1-N^c} \qquad (4.28)$$

where N^c and M^c are the demagnetizing field coefficients and the magnetization along the **c**-axis, respectively. Fig. 4.3 shows the variation of the magnetization of the single crystal as a function of the internal field. This experiment was carried out at very low temperature ($T = 1.5K$) with a very low rate of the variation of this field $dH/dt = 2 \times 10^{-3}T/min$. The deviation from the linearity of this curve marks the transition between the perfect and the partial diamagnetism which occurs at ~ 699 Oe for this single crystal.

The value of the demagnetizing field coefficient N^c can be determined in two ways: By the first method it is extracted from the slope of $M(H)$ whose equation in the Meissner state is $M^c = - H/4\pi(1 - N^c)$; In the second method, the sample is compared to a flattened ellipsoid of axes $a = b \neq c$, a simple reading in suitable abacuses provides the value of N^c.

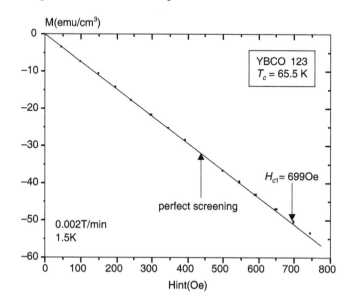

Figure 4.3. Magnetization as a function of the internal field of the single crystal $YBa_2Cu_3O_{7-\delta}$ for low values of the magnetic field. The deviation from the linearity marks the transition between the Meissner state and the mixed state (after S. Khene, 1999[4]).

The two methods are equivalent for a massive sample and lead more or less to the same result. In this case, $N^c = 0.8$. Knowing that at very low temperatures the lower field changes very little with the temperature $H_{C1}^c(1.5K) \sim H_{C1}^c(0)$, the value at very low temperature of the penetration depth, derived from Equation 4.17 by taking κ^c of the order of 100 and $\Phi_0 = 2.07 \ 10^{-7} G/cm^2$, is $\lambda^c(0) \sim 1040 \text{Å}$. This value is close to what is generally reported in the literature[5,6]. Strictly speaking, the value of the lower critical field depends on several parameters including the surface quality of the sample[7].

1.6. Anisotropy of the Magnetization

1.6.1. Magnetization Near Hc_2

For a value of the applied field H such as $H_{C2} - H \ll H_{C2}$, the two components of the magnetization, the parallel component $M_{//}(B,\theta_B)$ and perpendicular $M_\perp(B,\theta_B)$ component to the direction of the vortices, are given to the first order in $[1 - B/H_{C2}(\theta_B)]$ by[8]:

$$M_{//}(B,\theta_B) = -M_0 \frac{\varepsilon(\theta_B)}{\beta_A} \left[1 - \frac{B}{H_{C2}(\theta_B)} \right] \tag{4.29}$$

$$M_\perp(B,\theta_B) = \\ -M_0 \frac{(\gamma^2 - 1) \sin(2\theta_B)}{2\beta_A \, \varepsilon(\theta_B)} \left[1 - \frac{B}{H_{C2}(\theta_B)} \right] \tag{4.30}$$

here, θ_B indicates the angle between the direction of the vortices and the c-axis, M_0 defines the magnetization unit of the Ginzburg-Landau theory and is given by:

$$M_0 = \frac{\sqrt{2} \, H_C(T) \kappa}{8\pi} \tag{4.31}$$

and H_{C2} represents the geometric average:

$$H_{C2} = (H_{C2//}^2 \, H_{C2\perp})^{1/3} \tag{4.32}$$

1.6.2. Magnetization Within the Limit of London

Within the limit of London ($H_{C1} \ll H \ll H_{C2}$), the components of the magnetization are written[9-11]:

$$M_{//}(B,\theta_B) = -M_0 \, \varepsilon(\theta_B) \, \ln \sqrt{\frac{\eta \, H_{C2}(\theta_B)}{B}} \tag{4.33}$$

$$M_\perp(B,\theta_B) = \\ -M_0 \frac{(\gamma^2 - 1) \sin(2\theta_B)}{2\varepsilon(\theta_B)} \ln \sqrt{\frac{\eta \, H_{C2}(\theta_B)}{B}} \tag{4.34}$$

where $\eta \sim 1.15$ for a triangular lattice.

1.6.3. Magnetization in the Interval $H_{C1} \ll H < H_{C2}$

By taking into account the contribution of the core of the vortices, Z. Hao and J.R. Clem[12] propose the following expressions for the magnetization, valid for the entire field interval $H_{C1} \ll H < H_{C2}$:

$$M_{//}(B,\theta_B) = -\alpha \, M_0 \, \varepsilon(\theta_B) \, \ln \left[\frac{\beta \, H_{C2}(\theta_B)}{B} \right] \tag{4.35}$$

$$M_{\perp}(B,\theta_B) = -\,\alpha\, M_0\, \frac{(\gamma^2 - 1)\, sin\,(2\,\theta_B)}{2\,\varepsilon(\theta_B)} \quad (4.36)$$

$$\ln\!\left[\frac{\beta H_{C2}(\gamma_B)}{B}\right]$$

where α and β are two coefficients which weakly depend on the field and the temperature. For example for:

$$0.02 \leq \frac{H}{H_{C2}} \leq 0.1 \qquad \alpha \sim 0.84 \text{ et } \beta \sim 1.08$$

$$0.1 \leq \frac{H}{H_{C2}} \leq 0.3 \qquad \alpha \sim 0.70 \text{ et } \beta \sim 1.74 \qquad (4.37)$$

$$\frac{H}{H_{C2}} \sim 1 \qquad \alpha \sim 0.862 \text{ et } \beta \sim 1$$

In general, the angle θ between the applied field **H** and the **c**-axis of the crystal is different from the angle θ_B between the flux density **B** and the **c**-axis. This difference creates a transverse magnetization and a magnetic torque. Within the limit of London, this difference is given by:

$$sin\,(\theta - \theta_B) =$$

$$\frac{H_{C1\perp}\,(\gamma^2 - 1)}{H}\; \frac{sin\,\theta_B\, cos\,\theta_B}{\varepsilon(\theta_B)}\; \frac{\ln\!\left[\dfrac{H_{C2}(\theta_B)}{B}\right]}{2\ln\kappa_{\perp}} \qquad (4.38)$$

This equation shows that for $H \gg H_{C1\perp}$, one has $\theta \sim \theta_B$.

1.7. Structure of the Vortices Lattice

In the Shubnikov phase, the structure of the vortices lattice of an anisotropic superconductor in the London limit depends on the orientation of the magnetic field **H** with respect to the **c**-axis of the crystal. Thus, for an applied magnetic field parallel to the **c**-axis, the vortices adopt the usual configuration of Abrikosov. Now, if the same field is applied parallel to the ab planes, the lattice of vortices becomes very compressed in the direction of the **c**-axis and much dilated in the direction parallel to the planes. The triangles of the unit-cell become isosceles. For example, by assuming a field parallel to the **b**-axis, the ratio between the height of the triangle d_h and the distance d_c between vortices along the **c**-axis is of the order of[6]:

$$\frac{d_h}{d_c} \sim \frac{\lambda^c}{\lambda^{ab}} \qquad (4.39)$$

Between these two extreme cases, i.e. for a field which forms an angle θ with the c-axis, the unit-cell of the vortices lattice perpendicular to this lattice is defined by[9] (see Fig. 4.4):

$$\frac{b_2}{b_1} = \frac{1}{2\cos\beta} \ , \quad tan\,\beta = \frac{\sqrt{3}}{\varepsilon(\theta)} \tag{4.40}$$

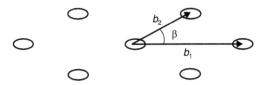

Figure 4.4. Transverse section of the unit-cell of the vortices lattice in an anisotropic superconductor when the applied field H makes a nonzero angle θ with the c-axis of the crystal.

1.8.   *Transverse Component of the Magnetization*

In the anisotropic superconductors, the magnetization **M**, the flux density **B** and the applied field **H** are not collinear. This situation leads to the appearance of a component of the magnetization, $\mathbf{M}_T$, perpendicular to the applied field which results in a torque Γ:

$$\Gamma = M_T \times H \tag{4.41}$$

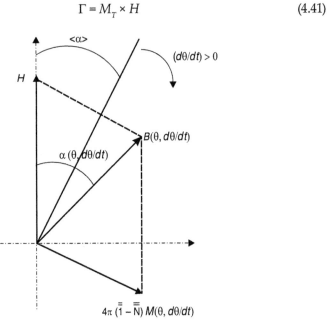

Figure 4.5. Graphical representation of Equation 4.42. The magnetic field lags behind the vortices.

Two methods permit the achievement of these rotational measurements. The first method consists in fixing the sample and in rotating the source of the magnetic field. The vortices in the sample lag behind the field. The second method consists in fixing the source of the field and in rotating the sample. This time, the magnetic lags behind the vortices. In the absence of the vortices pinning and in the limit where the sample size is much greater than the London penetration depth, the flux density in the material is given by:

$$\mathbf{B} = \mathbf{H} + 4\pi \left(\overline{\overline{1}} - \overline{\overline{N}} \right) \mathbf{M} \tag{4.42}$$

whose the projection according to the directions parallel and perpendicular to the applied field gives[13] (see Fig. 4.5):

$$4\pi(1 - N_L)\, M_L(\theta,\, d\theta/dt) = B_L(\theta,\, d\theta/dt)\, cos(\theta,\, d\theta/dt) - H \tag{4.43}$$

$$4\pi(1 - N_T)\, M_T(\theta,\, d\theta/dt) = B_T(\theta,\, d\theta/dt)\, sin(\theta,\, d\theta/dt) \tag{4.44}$$

where N_L and N_T, B_L and B_T, M_L and M_T denote the coefficients of demagnetizing fields, the flux densities and the magnetizations in the directions parallel and perpendicular to the applied field, respectively. In a high-T_C superconductor, the magnetic field does not drag at a constant angle behind the vortices because these last oscillate around an average angle $<\alpha(\theta)>$ leading to a periodic variation of the two components of the magnetization (see Fig. 4.6a and b). This oscillation is the result of a competition between three effects, the anisotropy of the material, the intrinsic and extrinsic vortices trapping and the restoratory strength of the magnetic field. For example, one associates the very acute extrema of $M_T(\theta)$ of Fig. 4.6b with the trapping of vortices by the lamellar structure of the material[13]. These measurements are very useful because they allow drawing very important information on the vortices configuration and in getting moderately anisotropic compounds all of their phenomenological parameters. This is the case for example of the $YBa_2Cu_3O_{7-\delta}$ compound which can be described by the anisotropic Ginzburg-Landau theory. If one takes the precaution to apply a field H higher than H_{C1}, it is possible to make, for example in the case of the transverse component, the following approximation $M_T(\theta) \sim M_T(B, \theta_B)$, and Equations 4.30 and 4.36 change into:

$$M_T(H, \theta) =$$

$$- M_0\, \frac{(\gamma^2 - 1)\, sin(2\theta)}{2\beta_A\, \varepsilon(\theta)} \left[1 - \frac{H}{H_{C2}(\theta)} \right] \tag{4.45}$$

$$M_T(H,\theta) =$$
$$-\alpha M_0 \frac{(\gamma^2 - 1)\,\sin\,(2\theta)}{2\varepsilon(\theta)}\,\ln\left[\frac{\beta H_{C2}(\theta)}{H}\right] \qquad (4.46)$$

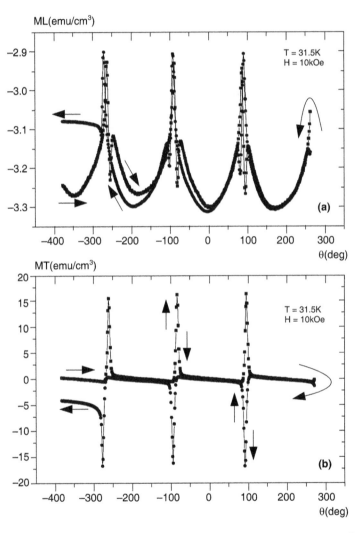

Figure 4.6. Angular variation of the magnetization of a single crystal of the type La$_{1.85}$Sr$_{0.15}$CuO$_4$ (a) Longitudinal component M_L (b) Transverse component M_T (after S. Khene, 1999[4a]).

If one has a magnetometer with double squid in perpendicular geometry and with a system which permits to rotate the sample in the two senses (see Fig. 4.7), it is possible to simultaneously measure the two components, the transverse component M_T and longitudinal component M_L, of the magnetization **M** and to extract the reversible part.

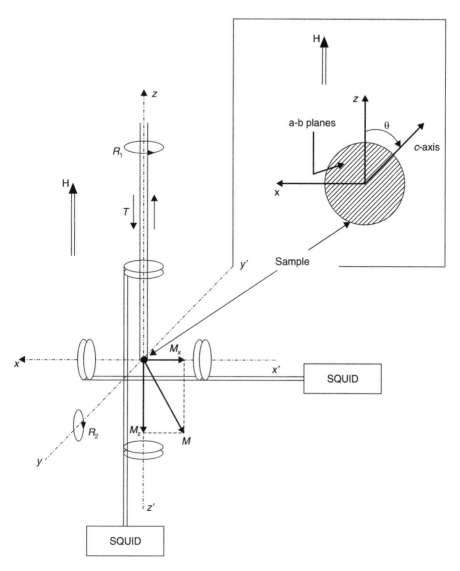

Figure 4.7. Magnetometer with two squids in perpendicular geometry and representation of the three degrees of liberty of the sample: a translation movement T along the z-axis, a rotational movement R_1 around the z-axis and a rotational movement around the y-axis that allows varying the angle θ between the direction of the applied field H and the c-axis of the crystal (see the insert) (after S. Khene, 2004[4b]).

The ratio of the two reversible components gives as well in the London model as close to H_{C2}[13,14]:

$$\frac{M_{Trev}}{M_{Lrev}} = (\gamma^2 - 1)\, \frac{\sin\theta\,\cos\theta}{\sin^2\theta + \gamma^2\,\cos^2\theta} \qquad (4.47)$$

Fig. 4.8 gives the angular variation of the ratio of the two components.

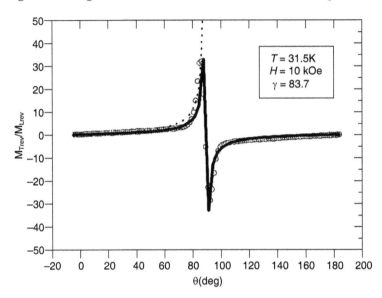

Figure 4.8. Angular variation of the ratio M_{Trev}/M_{Lrev} of the $La_{1.85}Sr_{0.15}CuO_4$ compound at $T = 31.5K$ and for an applied field of 10 kOe. The function tan θ is also shown in dashed lines (after S. Khene, 2004[4b]).

By taking into account the measurement errors, these curves are superimposed. The best fit obtained with the expression 4.47 gives an average value of γ equal to 83.7. This shows that this system is anisotropic. Fig. 4.8 also shows that one has:

$$\frac{M_{Trev}}{M_{Lrev}} \sim tan\ \theta \tag{4.48}$$

This result indicates that the M_{rev} vector is aligned along the **c**-axis during the rotation of the crystal. The study of the angular variation of the transverse component M_{Trev} for several values of the applied field allows obtaining the values of the maxima M_{Tmax} of each curve. A linear growth of the maximum of $(M_T)^{2/3}$ with the applied field is expected in accordance with Equation 4.49 which predicts the following dependence when H approaches the value $H_{C2}(T)/\varepsilon(\theta_{max})$:

$$M_T(\theta_{max}) = \frac{M_0\sqrt{\gamma^2 - 1}\left[\dfrac{1 - H\,\varepsilon(\theta_{max})}{HC_2^{ab}}\right]^{3/2}}{\beta_A} \tag{4.49}$$

It is then possible to extract the values of H_{C2}^{ab} and κ^{ab}.

2. Lawrence-Doniach Model

2.1. Josephson Effects

2.1.1. Continuous Josephson Effect

In 1962, B.D. Josephson[15] theoretically foresaw that if two superconductors *A* and *B* are separated by a very thin insulating layer, of the order of 10Å, a continuous current will cross the barrier without any applied voltage, the resistance staying of course zero. This is the continuous Josephson Effect. This current is born of the phase difference of the macroscopic wave function of the Cooper pairs in the superconductors *A* and *B*. It is expressed as:

$$I = I_g \, sin \, \gamma \tag{4.50}$$

where $\gamma = \varphi_A - \varphi_B$, it represents the phase difference through the junction. I_g is the maximum value of the current. This current is proportional to the surface of the junction; it exponentially decreases with its thickness. The current of the Cooper pairs cannot exceed I_g. Indeed, if one imposes a current $I > I_g$, the potential difference goes up to V_g. Beyond V_g, the junction becomes resistive; the current will be then composed of normal carriers (dissociated pairs) and becomes approximately linear in *V*. Theoretical considerations connect I_g to $\Delta(T)$, the superconducting gap (itself depending on the temperature)[16]:

$$I_g = \frac{\pi \, \Delta(T)}{2 \, e \, R_n} \, tanh \left[\frac{2\Delta(T)}{2 \, k_B \, T} \right] \tag{4.51}$$

where R_n is the resistance of the junction in the normal state. This equation, called "Ambegaokar-Baratoff equation", teaches us that $I_g \rightarrow 0$ at T_C as $(T_C - T)$ because $\Delta^2(T)$ is proportional to $|\psi|^2$, square of the order parameter of Ginzburg-Landau, itself proportional to $\alpha(T) = \alpha_0 \, (T_C - T)$. A $T = 0$K, one has:

$$I_g = \frac{\pi \, \Delta}{2 \, e \, R_n} \tag{4.52}$$

In a junction Superconductor/Normal metal/Superconductor (S/N/S), P.G. De Gennes[17] shows that $I_g \rightarrow 0$ at T_C like $(T_C - T)^2$. When a magnetic field is applied parallel to the plane of the junction, the surface currents appear to expel it from the material. As the limiting density of the junction I_g is lower than the critical current density of a type II superconductor ($I_g \sim 10^4$A/m² for an oxide junction of 1mm² of surface, $R_n \sim 0.1 \, \Omega$, $\Delta \sim 1$meV for a conventional superconductor[6]), the magnetic field penetrates the junction over distances, λ_J greater than the superconductor:

$$\lambda_J = \sqrt{\frac{c\Phi_0}{16\pi\,\lambda\,I_g}} \qquad (4.53)$$

λ_J is typically of the order of 1 mm whereas that $\lambda \sim 10^{-3}$mm. For the high values of the applied field H:

$$H > H_{C1}{}^J = \frac{\Phi_0}{4\pi\,\lambda_J^2} \qquad (4.54)$$

the so-called Josephson vortices appear in the junction. These vortices are distinguished from the Abrikosov vortices by the fact that they do not have a normal core. In other words, the order parameter of the two superconducting electrodes does not vanish at the center of this kind of vortices.

2.1.2. Alternative Josephson Effect

A current $I > I_g$ can therefore circulate even for a zero voltage. But if one imposes a finite voltage V, B.D. Josephson shows that the phase difference γ varies as:

$$\frac{d\gamma}{dt} = \frac{2\pi}{h} \times Energie = \frac{2\pi}{h} \times 2eV \qquad (4.55)$$

The current oscillates at a frequency:

$$\nu = \frac{2e}{h} \times V = \frac{V}{\Phi_0} \qquad (4.56)$$

The ratio ν/V depends only on the fundamental constants. It is equal to 484 MHz/μV. This is the alternative Josephson Effect. These two effects were experimentally confirmed. They are at the origin of phenomena of a great scientific interest and great importance for the applications[18].

2.2. Free Energy

In some high-T_C superconductors where the coherence length along the c-axis is much smaller than the periodicity of the crystal $\xi^c(T) \ll d$ (BiSCaCuO, for example), the anisotropic model of Ginzburg-Landau is not appropriate to describe the phenomena which occur there because it does not take into account the real heterogeneity of these compounds. This model is then replaced by a more general model: the Lawrence-Doniach model. This last applies to all the superconductors which can be modeled by an alternating superposition of superconducting and insulating layers (or metallic layers). Thereby, the discovery in 1970 by F.R. Gamble et al.[19] of the lamellar superconductors of the type TaS_2 (Pyridine)$_{1/2}$ formed of the metal layers separated by the layers of organic molecules, and the availability of experimental data on their properties lead W.E. Lawrence and S. Doniach[20] to propose in 1971 a free energy function, sum of the free energy of Ginzburg-Landau for the planes and the Josephson coupling energy between the planes. The order parameter has a nonzero value only in the superconducting layers (see Fig. 2.9):

$$
\begin{aligned}
F_{LD} = d \sum_n \int d^2\mathbf{r} \Bigg\{ &\alpha \left|\psi_n(\mathbf{r})\right|^2 + \frac{\beta}{2}\left|\psi_n(\mathbf{r})\right|^4 + \\
&\frac{1}{2m}\left|\left(-i\hbar\nabla_{/\!/} + \frac{e\mathbf{A}_{/\!/}}{c}\right)\psi_n\right|^2 \\
&+ J\left|\psi_n - \psi_{n+1}\ exp\left(-\frac{2\pi i}{\Phi_0}\int_{nd}^{(n+1)d} A_z\ dz\right)\right|^2 \Bigg\} + \\
&\int d^3 r\ \frac{h^2}{8\pi} - \frac{\mathbf{h}.\mathbf{H}}{4\pi}
\end{aligned}
$$

(4.57)

where J is the coupling coefficient, m the effective mass of the electron along the superconducting layers and where the equidistant superconducting planes are indexed by n and are supposed sufficiently thin ($d_0 \ll d$) so that the order parameter $\psi_n(x,y)$ can be considered independent of the z coordinate while changing continuously on each superconducting plane. The order parameter that appears in Equation 4.57 is in fact defined according to the order parameter of the superconducting layers ψ_{n0} by:

$$
\left|\psi_n\right|^2 = \frac{d_0}{d}\left|\psi_{n0}\right|^2
$$

(4.58)

The last definition allows, on the one hand homogeneously summing up all the quantities over the entire superconductor and on the other hand immediately going to the continuous limit by approaching d to zero.

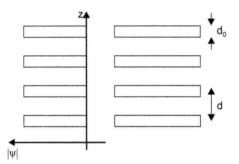

Figure 4.9. Representation of a laminar superconductor formed of a stack of superconducting layers of thicknesses d_0, separated by insulating layers. The order parameter has a nonzero value only in the superconducting layers.

The quantity $d\psi_n^2 = d_0 \psi_{n0}^2$ is proportional to the volume density n_s of the Cooper pairs in the material. The minimization of the function 4.57 compared to ψ_n and to the potential vector **A** allows obtaining the equations of Lawrence-Doniach (with $\tau = \hbar^2/2m$):

$$
\psi_n \left(\alpha + \beta |\psi_n|^2 \right) - \tau \left(\nabla_{//} - \frac{2i\pi}{\Phi_0} A_{//} \right)^2 \psi_n
$$

$$
+ J \left[2\psi_n - \psi_{n+1} \, exp \left(-\frac{2I\pi}{\Phi_0} \int\limits_{nd}^{(n+1)d} A_z \, dz \right) \right.
$$ (4.59)

$$
\left. - \psi_{n-1} \, exp \left(-\frac{2i\pi}{\Phi} \int\limits_{nd}^{(n-1)d} A_z \, dz \right) \right] = 0
$$

$$
\frac{1}{c} j_{//}(n) = \frac{1}{4\pi} (\nabla \times h)_{//}(n)
$$

$$
= \frac{2i\pi\tau}{\Phi_0} (\psi_n \nabla_{//} \psi_n^* - \psi_n^* \nabla_{//} \psi)
$$ (4.60a)

$$
- 2\tau \left(\frac{2\pi}{\Phi_0} \right)^2 |\psi_n|^2 A_{//}
$$

$$
\frac{1}{c} j_z(n) = \frac{1}{4\pi} (\nabla \times h)_z
$$

$$
= \frac{2\pi \, dJ}{\Phi_0} \cdot 2m \left[\psi_{n+1}^* \psi_n \, exp \left(\frac{2i\pi}{\Phi_0} \int\limits_{nd}^{(n+1)d} A_z \, dz \right) \right]
$$ (4.60b)

2.3. Characteristic Lengths

In the same way as the Ginzburg-Landau model, one introduces the following characteristic lengths:

$$\xi^{ab}(T) = \sqrt{\frac{\hbar^2}{2m\,|\alpha(T)|}} \quad ; \quad \xi^c(T) = \sqrt{\frac{J d^2}{|\alpha(T)|}} \tag{4.61}$$

$$\lambda^{ab}(T) = \sqrt{\frac{m c^2}{4\pi\, n_s\, e^2}} \quad ; \quad \lambda^c = \gamma\,\lambda^{ab} \tag{4.62}$$

The square of the anisotropy coefficient γ^2 is defined in this model as the ratio of the kinetic energy of a Cooper pair and the coupling energy between two adjacent superconducting layers:

$$\gamma^2 = \frac{\hbar^2}{2 m J d^2} \tag{4.63}$$

If one formally puts:

$$J = \frac{\hbar^2}{2 M d^2} \tag{4.64}$$

where M is the effective mass of the electrons in their movement by tunneling effect from a superconducting plane to another, one finds the anisotropy ratio of the anisotropic Ginzburg-Landau theory (see Equation 4.9):

$$\gamma = \sqrt{\frac{M}{m}} \tag{4.65}$$

The model of Lawrence-Doniach was justified at the microscopic level in 1973 by L.N. Bulaevskii[21]. This last connects, in the pure limit, the mass M to the integral of jump between the superconducting planes $t_\perp$ by the equation:

$$J = \frac{\hbar^2}{2 M d^2} = \frac{t_\perp^2}{4\varepsilon_F} \tag{4.66}$$

ε_F is the Fermi energy.

2.4. Change of the Dimensional Regime

2.4.1. Magnetic Field Parallel to the Layers

Equation 4.59 of Lawrence-Doniach permits determination of the upper critical field parallel to the layers $H_{C2}^{//}(T)$[20,22]. Thus, for a temperature close to T_c, one obtains:

$$H_{C2}^{//}(T) = \frac{\Phi_0}{2\pi \, \xi^{ab} \, \xi^c} \tag{4.67}$$

This critical field is identical to that given by the anisotropic Ginzburg-Landau theory, provided that one defines an effective coherence length $\xi^c(T)$ starting from the Josephson coupling (Eq. 4.61). In particular, this critical field varies as $[1 - (T/T_c)]$ near T_c. On the other hand, far from T_c, the resolution of the first-Lawrence Doniach equation leads to the following expression of $H_{C2}^{//}(T)$ which diverges at a certain temperature T^* (see Fig. 4.10):

$$H_{C2}^{//}(T) = \frac{\Phi_0}{\gamma d^2 \sqrt{1 - \frac{d^2}{2\xi_c^2(T)}}} \tag{4.68}$$

Indeed, as:

$$\xi^c(T) = \frac{\xi^c(0)}{\sqrt{1 - \frac{T}{T_C}}} \tag{4.69}$$

It can be seen that $H_{C2}^{//}$ varies as $(T - T_c)^{-1/2}$ for $T^* < T < T_c$, where:

$$T^* = T_C \left[1 - \frac{2\xi_c^2(0)}{d^2} \right] \tag{4.70}$$

The temperature T^* (or T_{Cr}) is called "change temperature of the dimensional regime". It separates two regimes: for $T < T^*$, the vortices do not have cores, a little like those found in the conventional Josephson junctions and for $T^* < T < T_c$, vortices are of the Abrikosov type[22]. These last are formed of an elliptic core of dimensions $\xi^{ab}(T)$ and $\xi^c(T)$ within which the order parameter is strongly decreased. Within this one, a non-linear elliptic core is also found, of sizes d and γd, due to the effects related to the spatial variation of the order parameter. Outside the core of the vortex, the tunnel screen current circulates and extends to a region bounded by lengths $\lambda^{ab}(T)$ and $\lambda^c(T)$[23-24] (see Fig. 4.11).

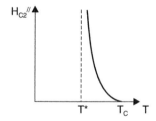

Figure 4.10. Critical field $H_{c2}^{\,/\!/}$ in the Lawrence-Doniach model as a function of the temperature. The $H_{c2/\!/}$ field is infinity for temperatures below the temperature T^*.

The structure of a vortex in parallel field was first studied at very low temperatures by J.P. Carton[23] and then extended to the quasi-3D system by D.Feinberg and A.M. Ettouhami[25-26]. This change of regime depends on the coherence length; it occurs for (see Fig. 4.12):

$$\xi^c(T^*) = \frac{d}{\sqrt{2}} \tag{4.71}$$

This condition is in fact satisfied if:

$$\xi^c(0) < \frac{d}{\sqrt{2}} \tag{4.72}$$

or if:

$$\frac{2\xi_{ab}^2(0)}{\gamma^2 d^2} < 1 \tag{4.73}$$

In the high-T_C superconductors where the coherence length is very short and the anisotropy ratio very large, this condition seems to be verified. In the highly anisotropic system, BiSCaCuO for example, one estimates $(T_C - T^*)$ unless 1K[25]. The divergence of $H_{c2}^{\,/\!/}(T)$ can also be explained in the infinitely thin superconducting layers model where the critical field $H_{c2}^{\,/\!/}$ of a layer is unlimited at all temperatures[25]. Indeed, for a thin layer of thickness d_0, this field is expressed as:

$$H_{c2}^{\,/\!/} = \frac{\Phi_0 \sqrt{3}}{\pi \xi d_0} \tag{4.74}$$

which diverges for infinitely thin layers where d_0 approaches 0. This is the orbital critical field because in practice $H_{c2}^{\,/\!/}$ always has a finite value because of the paramagnetic effects[27].

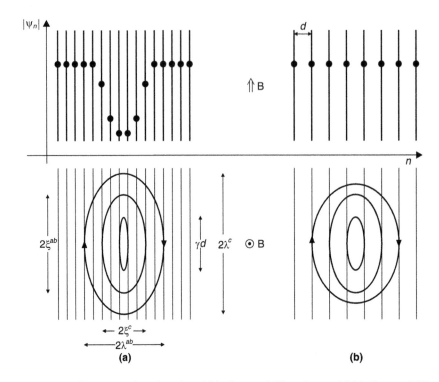

Figure 4.11. Representation of vortices (a) in the quasi-3D regime and (b) in the quasi-2D regime. With regard to the scales (not respected), it should be recalled that for the high-T_C oxides, the anisotropy parameter γ varies from 5 to about 1000 as a function of the compound and that $\kappa \approx 100$.

For $T < T^*$, the multilayer system is completely decoupled and the upper critical field is given by that of the individual layers which precisely diverges when the layers are reduced to zero thickness. The coherence length $\xi^c(T)$ has no sense and has no role with regard to the variations of the order parameter because in the expression of the free energy function of Lawrence-Doniach (see Eq. 4.57), it is implicitly assumed that $d_0 = 0$: the order parameter can then be considered constant in the z direction of the different superconducting planes (see Fig. 4.11b). The physical system is marked by the two-dimensionality. This regime is known as "quasi-2D Lawrence Doniach regime". For $T^* < T < T_C$, the upper critical field $H_{C2}^{//}$ has a finite value. This means that for this temperature range, the correlation between the superconducting layers is sufficiently strong so that the superconductors behave like an almost three-dimensional system; this system is called "quasi-3D system of the Lawrence-Doniach model". Although one can define in this regime a coherence length perpendicular to the superconducting planes $\xi^c(T) \geq d$, this does not mean that the order parameter continuously varies over the distance ξ^c in the z

direction. Indeed ψ_n is always defined only on the superconducting planes. ξ^c is to be interpreted as an effective coherence length which describes the macroscopic scale of the variation of the order parameter in the presence of the lamellar structure (see Fig. 4.11a).

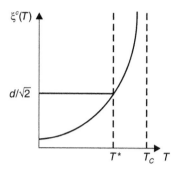

Figure 4.12. Variation of the coherence length ξ_c as a function of the temperature for a layered superconductor. Below T^*, $\xi_c (T)$ (= $\xi_{ab} (T)/\gamma$) does not have a very clear sense in the model of Lawrence-Doniach.

2.4.2. Magnetic Field Perpendicular to the Layers

For $T < T^*$, and an applied magnetic field **H** perpendicular to the layers, the vortices adopt a triangular configuration in each of the superconducting planes, the current loops will circulate only in these planes. These vortices which stack like plates along the H direction are called "2D vortex or pancakes"[28-33] (see Fig. 4.13).

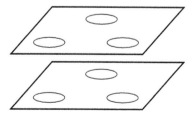

Figure 4.13. Triangular lattice of the 2D vortices in a layered superconductor when the applied field is perpendicular to the superconducting layers.

This is in fact a limiting case where the Josephson coupling is assumed to be zero, not to confuse with the true two-dimensional superconductivity of the very thin films because if in our case the electronic coupling is effectively of the type 2D, the magnetic coupling is of the type 3D. In addition, the pancakes of the same superconducting layer are pushed back whereas those of the neighboring layers attract each other.[32-33] In the quasi-

3D system, the configuration of vortices is very close to that which was described in the anisotropic Ginzburg-Landau theory.

2.4.3. Magnetic Field Inclined in Relation to the Layers

The structure of the vortices lattice in oblique field of the London version of the Lawrence-Doniach model was investigated in 1992 by D. Feinberg[34] and by L.N. Bulaevskii et al.[35] This structure in stairs depends on the angle θ_B between the direction of the vortices and the **c**-axis. Thus:

- For $\tan \theta_B < 2 \; \xi^{ab}/d$, the cores of the two 2D vortex belonging to the same flux line and located on two adjacent planes overlap. These flux lines are formed of slanted lines of 2D cores separated by broad regions where the London anisotropic currents circulate.

- For $\tan \theta_B > 2 \, \xi^{ab}/d$, the vortices cores on two neighboring planes are spaced at a distance $l = d \tan \theta_B > 2 \, \xi^{ab}$, and the Josephson vortex segments appear. One distinguishes in this case two sub regimes ($H_0 = \Phi_0/2\pi\gamma d^2$):

  - For $\tan \theta_B \ll \gamma \, (H_0/B)^{1/2}$, there are wide regions between vortices where the anisotropic London currents circulate.

  - For $\tan \theta_B \gg \gamma \, (H_0/B)^{1/2}$, the vortices are so close to each other that the regions where the current of London circulates, disappear.

REFERENCES

1 V.L. Ginzburg, Zh. Eksp. Teor. Fiz. **23**, 236 (1952).

2 A.V. Balatskii, L.I. Burlachkov and L.P. Gor'kov, Sov. Phys. JETP **63**, 866 (1986).

3 R.A. Klemm and J.R. Clem, Phys. Rev. **B 21**, 1868 (1980); R.A. Klemm, Phys. Rev. **B 41**, 117 (1980).

4 (a) S. Khène, Thesis, Badji Mokhtar University of Annaba, Algeria, 1999; (b) S. Khene, Physica B **349** (2004) 227–237.

5 G.V.S. Rao and U.V. Varadaraju, Chemistry of High Temperature Superconductors, ed. C.N.R. Rao, World Scientific, Singapore, 1991.

6 M. Cyrot and D. Pavuna, Introduction to Superconductivity and High-T_c Materials, World Scientific, 1992.

7 M. Konczykowski, L.I. Burlachkov, Y. Yeshurun and F. Holtzberg, Rev. **B 43**, 13707 (1991); L.I. Burlachkov, M. Konczykowski, Y. Yeshurun and F. Holtzberg, J. Appl. Phys. **70**, 5759 (1991).

8 V.G. Kogan and J.R. Clem, Phys. Rev. **B 24**, 2497 (1981).

9 L.J. Campbell, M.M. Doria and V.G. Kogan, Phys. Rev. **B 38**, 2439 (1988).

10 A.I. Buzdin and D. Feinberg, Phys. Lett. **A 165**, 281 (1992).

11 G. Blatter, V.B. Geshkenbein and A.I. Larkin, Phys. Rev. Lett. **68**, 875 (1992).

12 Z. Hao and J.R. Clem, Phys. Rev. **B 43**, 7622 (1991); Z. Hao and J.R. Clem, Phys. Rev. Lett. **67**, 2371 (1991).

13 P. Pugnat, Thesis, Joseph Fourier University of Grenoble I, France, 1995.

14 M. Tuominen, A.M. Goldman, Y.Z. Chang and P.Z. Jiang, Phys. Rev. **B 42**, 412 (1990).

15 B.D. Josephson, Phys. Rev. Lett. **1**, 251 (1962); Rev. Mod. Phys. **36**, 216 (1964).

16 V. Ambegaokar and A. Baratoff, Phys. Rev. Lett. **10**, 486; **11**, 104 (erratum) (1963).

17 P.G. De Gennes, Reviews of Mod. Phys. **36**, 225 (1964); Superconductivity of Metals and Alloys, W.A. Benjamin, New York, 1966.

18 K.K. Likharev, Rev. Mod. Phys. **51**, 101 (1979).
19 F.R. Gamble, F.J. Di Salvo, R.A. Klemm and T.H. Geballe, Science **168**, 568 (1970).
20 W.E. Lawrence and S. Doniach, Proc. of the 12th Conf. On Low Temp. Phys., Kyoto, E. Kanda ed., p. 361, 1971.
21 L.N. Bulaevskii, Sov. Phys. JETP **37**, 1133 (1973).
22 R.A. Klemm, A. Luther and M.R. Beasley, Phys. Rev. **B 12**, 877 (1975).
23 J.P. Carton, J. Phys. I (Condensed Matter) **1**, 113 (1991).
24 J.R. Clem, M.W. Coffey and Z. Hao, Phys. Rev. **B 44**, 2732 (1991).
25 A.M. Ettouhami, Thesis, Joseph Fourier University of Grenoble I, 1994.
26 D. Feinberg and A.M. Ettouhami, Int. Jour. of Mod. Phys. **B 7**, 2085 (1993).
27 A.M. Clogston, Phys. Rev. Lett. **9**, 266 (1962).
28 K.B. Efetov, Sov. Phys. JETP **49**, 905 (1979).
29 A. Buzdin and D. Feinberg, J. Phys. France **51**, 1971 (1990).
30 F. Guinea, Phys. Rev. **B 42**, 6244 (1990).
31 K.H. Fisher, Physica **C 178**, 161 (1991).
32 M.V. Feigelman, V.B. Geshkenbein and A.I. Larkin, Physica **C 167**, 177 (1990).
33 J.R. Clem, Phys. Rev. **B 43**, 7837 (1991).
34 D. Feinberg, Physica **C 194**, 126 (1992).
35 L.N. Bulaevskii, M. Ledvij and V.G. Kogan, Phys. Rev. **B 46**, 366 (1992).

Chapter 5

Dynamic of Vortices

1. Hysteresis Origin in the Magnetization Curves

For an applied field between H_{C1} and H_{C2}, the magnetic flux penetrates the superconductor in the form of vortices. The magnetization measurements $M(H)$ show that the curves in increasing fields never coincide with those obtained in decreasing fields (*see* Fig. 5.1). This is because the vortices have difficulties in moving as well as when they penetrate the material as when they leave. The origin of this phenomenon is the existence of defects which trap the vortices and thus allow the superconductor to support very high densities of current without a loss of energy. The critical current density is the extrinsic current quantity that liberates the vortices. Among the defects which can exist in the material let us quote the chemical defects such as the lacuna and the substitutions, and the structural defects such as the dislocations and the grains boundaries. The pinning effects are the result of three types of interaction, the vortex-defect interaction, the vortex-vortex interaction and the vortex-spin interaction.

2. Breaking Current of Cooper Pairs

Like the thermodynamic field H_C, the breaking Cooper pairs current is an intrinsic value of the superconductor. It represents the maximal theoretical value of the current that the material can support. This current breaks the Cooper pairs and consequently destroys the superconductivity. In the London model, its expression is obtained by equaling the kinetic energy of the current to condensation energy, so:

$$J_d = \frac{cH_C}{\lambda} \tag{5.1}$$

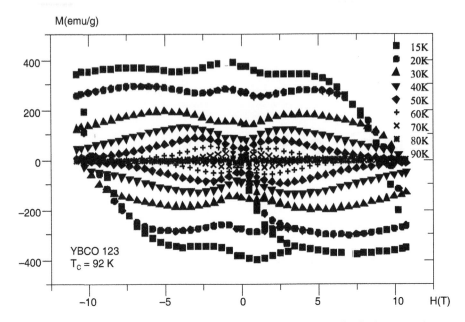

M(emu/g)

Figure 5.1. Magnetization curves $M(H)$ of a single crystal of $YBa_2Cu_3O_{7-\delta}$ for a speed $dH/dt = 0.47T/min$ with H parallel to the c-axis (after S. Khene, 1999[1]).

Its value is about $10^8 A/cm^2$ in conventional superconductors and about $10^9 A/cm^2$ in the high-T_C superconductors. In practice, these values are never reached. For the conventional superconductors, the highest observed value is the order of 10^6 to $10^7 A/cm^2$ [2] whereas for the high-T_C superconductors, it is about $10^7 A/cm^2$ in the best thin layers of the $YBa_2Cu_3O_{7-\delta}$ system[3] and about $10^6 A/cm^2$ [1a] in single crystals of the same system at very low temperature.

3. Pinning Force

The Lorentz force per unit volume which acts on an isolated vortex in the presence of a transport current **J** is given by:

$$\mathbf{f} = \mathbf{J} \times \Phi_0 \tag{5.2}$$

The force per unit volume which acts on the vortices lattice is obtained by multiplying the equation 5.2 by the density of vortices per unit surface $n = B/\Phi_0$, so:

$$\mathbf{F} = \mathbf{J} \times n \, \Phi_0 = \mathbf{J} \times \mathbf{B} \tag{5.3}$$

This force tends to displace the flux lines. If the vortices move freely, it is not possible to pass a very small current without energy loss. This means that to achieve high current densities, it is necessary to block the movement of the vortices. Let F_p be, the average density of the trapping force. It follows from what was said that if $F < F_p$, the vortices lattice remains motionless and one obtains a non-dissipative current. In the contrary case, that is to say for $F > F_p$, the vortices begin to move, and one then speaks about the flux flow regime. For $F = F_p$, one obtains the critical state from where one defines the critical current density J_c at $T = 0$K by:

$$F_p = J_c B \qquad (5.4)$$

The density of critical current J_c strongly depends on the irreversibility. It is zero for an ideal material (no defects). Based on the value of the transport current density J with respect to this density of the critical current J_c, one defines four different regimes, the flux flow regime (FF), the critical state, the flux creep regime (FC) and the thermally activated flux flow regime (TAFF).

4. Flux Flow

This mechanism appears when the density of the external applied current J exceeds the value of the critical current density $(J > J_c)$. The vortices lattice then starts moving and the resistivity of the material becomes linear in B. The energy loss which accompanies this movement can be described in terms of viscosity.[4] Therefore, the driving speed of the lines of flux v is obtained as a function of the viscosity η by simply equalizing the Lorentz force to the friction force:

$$J \, \Phi_0 = \eta \, \mathbf{v} \qquad (5.5)$$

This vortices motion induces an electrical field **E** parallel to the current **J** whose expression is given by the Maxwell equation:

$$\mathbf{E} = \mathbf{B} \times \mathbf{V} \qquad (5.6)$$

By using the two previous relationships, one obtains the resistivity of the flux flow:

$$\rho_{FF} = \frac{E}{J} = B \, \frac{\Phi_0}{\eta} \qquad (5.7)$$

The viscosity η is unknown. Experimentally, it does not depend on the current and varies very little with B. A calculation at a microscopic scale connects the flux flow resistivity, ρ_{FF}, to the resistivity, ρ_n, of the normal state for which all the superconducting electrons are normal in the field and at the temperature of measurement:[6]

$$\rho_{FF} = \rho_n \frac{B}{B_{C2}} \tag{5.8}$$

This allows us to express the viscosity in terms of resistivity:

$$\eta = \frac{\Phi_0 B_{C2}}{\rho_n} \tag{5.9}$$

The resistivity in the flux flow regime is similar to that obtained for currents circulating within the normal cores of the vortices. If the vortex is assimilated to a metallic cylinder of radius ξ, the ratio B/B_{C2} then represents the fraction of this normal metal. Furthermore, the viscous training in a conventional Josephson junction was first studied by P. Lebwohl and M.J. Stephen;[5] J.R. Clem and M.W. Coffrey[6] subsequently determined the flux flow resistivity induced by the motion of the Josephson vortices between the ab plane in the high-T_C superconductors. The basic idea behind these calculations is to consider that the cores of the vortices are crossed by unpaired electrons. However, C. Caroli et al.[7] and J. Bardeen et al.[8] showed that this assumption is not valid at very low temperatures (of the order of 20 to 30K for the YBCO 123 system for example[9]).

5. Bean Model

5.1. Description

According to the paragraph 4, one knows that for $J > J_c$, in other words for $F > F_p$, it appears as a viscous movement of the vortices lattice; this movement will persist in the material until an equilibrium state, for which $F = F_p$ in all points, is reached. This new equilibrium state is called "critical state"; and is associated with a critical gradient in the distribution of the vortices.

In 1962 C.P. Bean[10] introduced a simple model based on this concept. This model is the basis for the interpretation of the irreversible phenomena of the type-II superconductors which appear in magnetization curves, in alternative susceptibility measurements and in the dependence in temperature of the zero fields cooled and fields cooled magnetization.[10-11] It assumes that the material shows an opposition to the flux variation due to the applied field by creating a magnetization M generated by

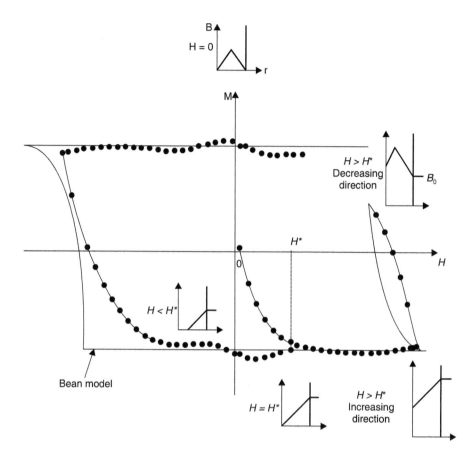

Figure 5.2. Magnetization process within the Bean model. The curve in dotted lines is experimental. It is that of a single crystal of YBa$_2$Cu$_3$O$_{7-\delta}$ carried out for a speed $dH/dt = 0.47$T/min, H parallel to the c-axis and T = 15K. B is the density of flux which prevails inside a cylindrical sample, r is a distance counted starting from its center. B_0 is the value of the density of flux at the sample/vacuum interface. For $H = H^*$, the vortices reach the center of the sample. The magnetization density is the area between B and B_0. In this model, the density of the current J is constant and equal to J_c in the regions penetrated by the magnetic flux, and zero elsewhere.

the current loops. The energic interest of the material is there then to oppose this applied field with the highest possible current density possible that is J_c. By assuming a J_c independent of the applied field, the magnetization process of a type-II superconductor in the framework of this model can be described as follows (see Fig. 5.2): when the value of the applied field reaches H_{C1}, the vortices which begin to penetrate the material from its surfaces are hindered by defects; in increasing fields, the number of vortices is therefore greater near the edges than in the center. For a particular value of the field H^*, the vortices reach the center of the sample, the magnetization is said to be saturated. For a single crystal of $YBa_2Cu_3O_{7-\delta}$, H^* is the order of 2 teslas at low temperatures whereas it is a thousand times lower in the case of an oriented powder. This shows that it is more difficult to saturate a single crystal.[12] Now, if one reduced the applied field the vortices would leave the sample, and the first to do it would be those that were near the surface. When the applied field returns to zero, a certain quantity of the magnetic flux remains blocked in the material because of the pinning; the vortices which are there are late compared to the decrease in the applied field. This explains the observed hysteresis in the magnetization curves. In the critical state concept, the heterogeneous distribution of the flux density B inside of the sample generates, according to the Maxwell-Ampere law, a current equal to J_C:

$$\nabla \times \mathbf{B} = \frac{4\pi}{c} J_C \tag{5.10}$$

The flux density $B(r)$ at the point r is obtained by merely integrating this last equation. The value of the measured magnetization is given by:

$$M(H) = \frac{1}{4\pi} \left[\frac{\int d^3r \; B(r)}{\int d^3r} - H \right] \tag{5.11}$$

Let us consider a type-II superconductor cooled in zero fields, to which one applies an increasing field parallel to its largest size. Two cases are presented for the magnetization curve:

$$H < H^* : \; 4\pi M(H) = -H + \alpha \frac{H^2}{H^*} - \beta \frac{H^3}{H^{*2}} \tag{5.12}$$

$$H > H^* : \; 4\pi M(H) = -\gamma H^* \tag{5.13}$$

where H^*, α, β and γ depend on the sample shape. Table 5.1 summarizes their values for two particular semi-infinite shapes of the sample, the cylinder of radius R and the plate of thickness D.

Table 5.1. Values of the factors H^*, α, β et γ for two particular shapes of the sample.

Factor	Cylinder of radius R	Plate of thickness D
H*	$(2p/c) \, J_c \, 2R$	$(2p/c) \, J_c \, D$
α	1	1/2
β	1/3	0
γ	1/3	1/2

Besides taking a J_C independent of the applied field H, the Bean model in its original version also assumes that the material is homogeneous and isotropic and there is no threshold field at the onset of the vortices ($H_{C1} = 0$).

This model does not take account of the thermal fluctuations ($T = 0$), the effects of the demagnetizing field ($N = 0$) and the reversible contribution to the magnetization. It is purely phenomenological. Its microscopic justification was nevertheless clarified by J. Friedel et al.[13] and by P.W. Anderson[14] who showed that the vortex undergoes the action of two forces, a magnetic pressure force exerted by the neighboring vortices and a pinning force due to the structural defects of the material.

5.2. Critical Current Evaluation

5.2.1. Isotropic Material

The critical current is generally estimated using the Bean model and the following relationships:[10, 15-1]

$$J_C(H) = 3 \frac{\Delta M}{a} \quad \text{or} \quad J_C(0) = 6 \frac{M_R}{a} \tag{5.14}$$

where $J_C(H)$ represents the critical current in a given field (in A/m²), ΔM the difference of the magnetizations in increasing and decreasing fields (in A/m) and « a » the side of the square ($a \times a$) perpendicular to the applied field (in m). $J_C(0)$ is the zero-field critical current obtained from the residual magnetization M_R. This relationship is strictly adapted to conventional superconductors but because of its simplicity, it continues to be used by many researchers in in the case of high-T_C oxides.

5.2.2. Anisotropic Material

The high-T_C superconductors are characterized by a strong structural anisotropy which leads us to distinguish three densities of critical current, $J_C^{c,ab}$, $J_C^{ab,ab}$ and $J_C^{ab,c}$.

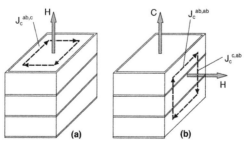

Figure 5.3. Representation of the three critical currents of an anisotropic superconductor. (a): $H \perp (a,b)$, (b): $H /\!/ (a,b)$. The first exponent indicates the direction of the current and the second that of the applied field.

The first exponent indicates the direction of the current and the second that of the applied field (see Fig. 5.3). By neglecting the demagnetizing field effects, for a parallel applied field to the c-axis of the crystal of rectangular section $(a_1 \times a_2)$ such as $a_1 > a_2$ one gets the following expressions:[17]

$$J_C(H) = \frac{2\,\Delta M}{a_2 \left(1 - \dfrac{a_2}{3a_1}\right)} \qquad \text{or}$$

$$J_C(0) = \frac{4\,M_R}{a_2 \left(1 - \dfrac{a_2}{3a_1}\right)} \tag{5.15}$$

6. Flux Creep

One recalls that for $T = 0K$, the vortices movement is only possible if the Lorentz force exceeds the pinning force leading to a current without loss of energy if $J < J_C(0, H)$. When the external current exceeds the critical current value, the vortices movement produces an electrical field and a tension: this is the flux flow regime described in the paragraph 4. However, for a finite temperature, there is a non-zero probability that the thermal agitation causes thermally activated transitions of the flux lines from a pinning potential to another creating a flux movement and thus a non-zero resistivity. This phenomenon appears even for $J < J_C(T, H)$. It is called "flux creep (or merely FC)". In order to explain the magnetic relaxation observed in the type-II conventional superconductors, P.W. Anderson[14] proposes in 1962 the flux creep theory based on the movement of the thermally activated flux. In this theory, the speed at which the flux bundles jump the pinning barriers U is given by the Arrhenius law:

$$v = v_0\, e^{-\frac{U(T, H, J_C)}{kT}} \tag{5.16}$$

where:

$$v_0 \sim v_p(T) + v_v(T) \tag{5.17}$$

with:

$$v_v(T) = c^2\, \rho_n(T) \left(\frac{B}{B_{C2}}\right)$$
$$\left(1 - \frac{B}{B_{C2}}\right)\left(\frac{1}{16\pi^2\, \lambda^2\,(T=0)}\right) \tag{5.18}$$

or:

$$v_v(T) = 6 \times 10^{10} \left(\frac{B}{B_{C2}} \right)$$

$$\left(1 - \frac{B}{B_{C2}} \right) (\rho_n^{-1}) \left[\frac{1000}{\lambda(0)} \right]^2 \tag{5.19}$$

where 6×10^{10} is in s^{-1}, ρ_n in $\mu\Omega cm^{-1}$ and 1000 in Å. Here $U(T,H,J_c)$ is the effective energy barrier which depends on T, H, the critical current and the direction of the flux jump in relation to the current density direction. Equation 5.17 is not unique, other combinations such as $v_0 \sim (v_p^2 + v_v^2)^{-1/2}$ are also possible.

7. Thermally Assisted Flux Flow

The notion of the thermally activated flux flow (TAFF) is a new terminology which appeared with the high-T_C oxides[18]. Its exact definition is not very clear and defers from author to author. It seems that this mechanism corresponds to the area of the T-H plane where the electrical field is proportional to J_C, the critical current density. In other words, the resistivity ($\rho = E/J$) is independent of J_C. This is compared with the limit of the usual creep where ρ varies exponentially with J_C. The TAAF mechanism has the same physical origin as the flux creep mechanism and consequently obeys to the same basic equation 5.16. It also satisfies the same equation of continuity for the magnetic induction B, equation that expresses the flux conservation. The main difference with the flux creep mechanism is that, in the TAAF mechanism, the reverse jump of the vortex cannot be overlooked. Let us note that the TAAF mechanism also exists in the conventional superconductors[19]. In the YBCO 123 system, this mechanism is observable only at $\sim$ 10K below T_C. This interval is larger for the bismuth compounds. In this part of the H-T plane, the variation of the irreversible magnetization as a function of the time follows a diffusion law of the type $M_{ir}(T) = M(t_0) \, exp \, (-t/\tau_{ff})$. With regard to the YBCO 123 system, the same exponential decrease was observed near T_C with an observation time of five days. This diffusion mechanism seems to occur for a very short observation time in the case of the bismuth or thallium compounds[20].

8. Irreversibility Line

Several experiments on the high-T_C superconductors can be explained by the TAAF mechanism and quantitatively studied by considering the flux movement as a linear diffusion. This allowed one to define in the H-T plane a line of irreversibility $T_{irr}(H)$ called "depinning line"[21-24], above which the vortices pinning disappears following the thermal depinning. Along of this depinning line, one has a constant diffusion of flux equivalent to $\rho(T,H) = C^{te}$, a phenomenon encountered in the eight following experiences [25]:

   i) The resistive transition enlargement $\rho(T,H)$ in a magnetic field.
   ii) The disappearance of the irreversibility in the magnetization curves at $T_{irr}(H)$.
   iii) The maximum in the imaginary part of the AC susceptibility $\mu(T,H,\omega) = \mu' + i\mu''$[18].
   iv) The divergence at $T_d(H)$ of the AC current penetration length measured by shielding of a AC current field by using a superconducting layer placed between two coils. This behavior was initially interpreted as a pair breaking line[26].
   v) The sharp maximum in the attenuation Γ of a vibrating high-T_C superconductor, mounted on silicon[27-29] or of a vibrating superconductor at $T_d(H)$[21-25,30-31].
   vi) The noise transmission in the high-T_C superconductors layers[32]. A sharp peak in the power of the noise at a given frequency according to H and T reproduces the melting line of the reference 27.
   vii) The ultrasonic attenuation peak and the increase of the sound speed[33].
   viii) The irreversibility in the width of the line depending on H and T in the muon rotation experiments (μ^+SR)[34].

9. Magnetic Instabilities

The flux jumps in the superconductors are the magnetic instabilities after which the magnetization disappears completely or partially under the effect of an avalanche of flux in the sample.

   These instabilities were observed in the type-II conventional superconductors[35-37] and interpreted[38] in terms of a thermodynamic disaster which occurs when the magnetic diffusion is larger than the thermal diffusion. In this case, the heat which is due to the vortices redistribution exceeds the capacity of the sample or its environment to absorb it; the vortices then start moving and cause a spontaneous destruction of the critical state of the magnetization[39]. It results in jumps of the magnetization between two limit curves corresponding to stable and unstable magnetic states. In most experiments, a detailed analysis of the flux jumps uses dynamic considerations with fundamental parameters such as the relative amplitudes of the thermal and magnetic diffusions. Nevertheless, the critical parameters which govern the possible occurrence of a flux jump can be merely deducted from the adiabatic theory of P.S. Swartz and C.P. Bean[38]. This theory predicts that a superconductor can be fully stable in relation to the flux jumps if a critical size of the sample is not exceeded[39]. This critical size depends on the critical current density, its derivative in relation to the temperature and the specific heat of the sample. This theory also predicts that the field of the first jump, and therefore

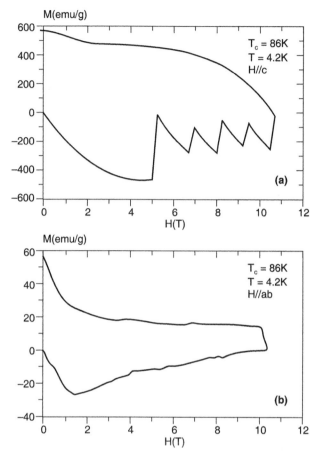

Figure 5.4. Magnetization of a single crystal of YBa$_2$Cu$_3$O$_{7-\delta}$, measured at $T = 4.2$K as a function of the applied magnetic field parallel (a) to the *c*-axis of the crystal (b) to the ab plane where the flux jumps entirely disappear (effect of the anisotropy of the crystal and the green phases present in small proportions in this sample) (after S. Khene and B. Barbara, 1999[41]).

the other jumps of flux, occurs in well-defined adiabatic conditions. These flux jumps were particularly studied in the type-II conventional superconductors in the form of cables or magnetic coils where their apparition was avoided by using multi-filamentary cables[40]. Because of the granularity of the high T_C ceramics which limits the dimension on which the screen currents circulate, the critical size predicted by the adiabatic theory is not exceeded and the flux jumps are therefore exceptionally observed in these materials. With the current possibility of producing large single crystals and polycrystals with big grains, the flux jumps are now observable in all families of the high-T_C oxides, Y-Ba-Cu-O[41,42-48] (see Fig. 5.4), Ho-Ba-Cu-O[49], Bi-Sr-Ca-Cu-O[50-51], La-Sr-Cu-O[52-53]. Studies carried out on these new materials show that the field of the first jump H_{fj} depends on several external parameters and the measurement conditions (speed and direction of sweep of the applied field, relaxation time, used apparatus, anisotropy, size of the sample, trapping mechanisms...)[41,42-43,46-,48-55], and that it cannot be adequately described within the framework of the adiabatic theory of Swartz and Bean. Several models were proposed in literature; unfortunately, these models do not simultaneously take into account all these parameters.

REFERENCES

1   S. Khene, Thesis, Badji Mokhtar University of Annaba, Algeria.
2   M. Cyrot and D. Pavuna, Introduction to Superconductivity and High-T_C Materials, World Scientific, 1992.
3   P. Chaudhari, R.H. Koch, R.B. Laibowitz, T.R. Mc Guire and R.J. Gambino, Phys. Rev. Lett. **58**, 2684 (1987).
4   J. Bardeen and M.J. Stephen, Phys. Rev. **140 A**, 1197 (1965).
5   P. Lebwohl and M.J. Stephen, Phys. Rev. **163**, 376 (1967).
6   J.R. Clem and M.W. Coffey, Phys. Rev. **B 42**, 6209 (1990).
7   C. Caroli, P.G. De Gennes and J. Matricon, J. Phys. Lett. **9**, 307 (1964).
8   J. Bardeen, R. Kümmel, A.E. Jacobs and L. Tewrdt, Phys. Rev. **187**, 556 (1969).
9   P. Martinoli, Ph. Flueckiger, V. Marsico, P.K. Srivastava, Ch. Leemann and J.L. Galivano, Physica **165-166B**, 1163 (1990).
10 C.P. Bean, Phys. Rev. Lett. **8**, 250 (1962); Rev. Mod. Phys. **36**, 31 (1964).
11 L. Krusin-Elbaum, A.P. Malozemoff, D.C. Cronemeyer, F. Holtzberg, J.R. Clem and Z. Hao, J. Appl. **67**, 4670 (1990).
12 B. Janossy, Thesis, University of Paris-South, France, 1993.
13 J. Friedel, P.G. De Gennes and J. Matricon, J. Appl. Lett. **2**, 119 (1963).
14 P.W. Anderson, Phys. Rev. Lett. **9**, 309a (1962).
15 P.W. Anderson and Y.B. Kim, Rev. Mod. Phys, **39** (1964).
16 M.R. Beasley, R. Labusch and W.W. Webb, Phys. Rev. **181**, 682 (1969).
17 G.W. Crabtree, J.Z. Liu, A. Unezawa, W.K. Kwok, C.H. Sowers, S.K. Malik, B.W. Veal, D.J. Lam, S.K. Brodsky and J.W. Downey, Phys. Rev. **B 36**, 4021 (1987).
18 P.H. Kes, J. Aartz, J. Van Den Berg, C. Jvan Den Beek and J.A. Mydosh, Supercond. Sci. Technol. **1**, 242 (1989).
19 P. Berghuis and P.H. Kes, Physica **B 165–166**, 1169 (1990).
20 D. Dew-Hughes, Proceedings Int. Conf. On Transp. Properties of Superconductors. Rio de Janeiro (Brazil), April-May 1990, (World Scientific, Singapore).
21 A. Gupta, P. Esquinazi, H.F. Braun, W. Gerhäuser, H.W. Neumüller, K. Heine and J. Tenbrink, Europhys. Lett. **10**, 663 (1989); A. Gupta, P. Esquinazi, H.F. Braun and E.H. Brandt, Physica **C 162-164**, 667 (1989).
22 P. Esquinazi, Solid State Commun. **74**, 75 (1990).
23 A. Gupta, P. Esquinazi and H.F. Braun, Phys. Rev. Lett. **63**, 1869 (1989).
24 P. Esquinazi, A. Gupta and H.F. Braun, Physica **B 165-166**, 1151 (1990).
25 E.H. Brandt, J. Mod. Phys. **B 5**, 751 (1991).
26 A. Hebard, P. Gammel, C. Rice and A. Levi, Phys. Rev. **B 40**, 5243 (1989).
27 P.L. Gammel, L.F. Schneemeyer, J.V. Waszczak and D.J. Bishop, Phys. Rev. Lett. **61**, 1666 (1988).
28 E.H. Brandt, P. Esquinazi and G. Weiss, Phys. Rev. Lett. **62**, 2330 (1989).
29 R.N. Kleimann, P.L. Gammel, L.F. Schneemeyer, J.V. Waszczak and D. Bishop, Phys. Rev. Lett. **62**, 2331 (1989).

30 E.H. Brandt, P. Esquinazi and H. Neckel, J. Low. Temp. Phys. **63**, 187 (1987); E.H. Brandt, P. Esquinazi, H. Neckel and G. Weiss, Phys. Rev. Lett. **56**, 89 (1986); E.H. Brandt, J. de Physique **C 8**, 31 (1987); P. Esquinazi, H. Neckel, E.H. Brandt and G. Weiss, J. Low temp. Phys. **64**, 1 (1986).

31 P. Esquinazi, C. Duran and E.H. Brandt, J. Appl. Phys. **65**, 4936 (1989).

32 A. Maeda, Y. Kato, H. Watanabe, I. Terasaki and K. Uchinokura, Physica **B 165–166**, 1363 (1990).

33 J. Pankert, Physica **C 168**, 335 (1990); J. Pankert, G. Marbach, A. Comberg, P. Lemmens, P. Frönig and S. Ewert, Phys. Rev. Lett. **65**, 3052 (1990); P. Lemmens, P. Frönig, S. Evert, J. Pankert, G. Marbach and A. Combey (cited by E.H. Brandt, J. Mod. Phys. **B 5**, 751 (1991).

34 P. Zimmermann, H. Keller, W. Kündig, B. Pümpin, I.M. Savié, J.W. Schneider and H. Simmler, Hyperfine Interactions **63–65**, 25 (1990).

35 B.B. Goodman and M. Wertheimer, Phys. Letters **18**, 236 (1965).

36 S.H. Goedeeemoed, C. Van Kolmeschate, J.W. Metselaar and D. De Klerk, Physica **31**, 573 (1965).

37 S.H. Goedemoed, C. Van Kolmeschate, P.H. Kes and D. De Klerk, Physica **32**, 1183 (1966).

38 P.S. Swartz and C.P. Bean, J. Appl. **39**, 4991 (1968).

39 S.L. Wipf and M.S. Lubell, Phys. Lett. **16**, 103 (1965); S.L. Wipf, Phys. Rev. **161**, 404 (1967).

40 M.N. Wilson, Superconducting Magnets, Clarendo Press, Oxford, 1983.

41 S. Khene and B. Barbara, Solid St. Commun. **109**, 727 (1999).

42 J.L. Tholence, H. Noël, J.C. Levet, M. Potel and P. Gougeon, Solid State Commun. **65**, 1131 (1988), S. Khene, Joint European Magnetic Symposia (JEMS'04), Dresden, Germany, September 05–10, 2004. S. Khene, JMMM 290–291, 981 (2005).

43 M. Guillot, M. Potel, P. Gougeon, H. Noël, J.C. Levet, G. Chouteau and J.L. Tholence, Phys. Lett. **A 127**, 363 (1988).

44 K. Chen, S.W. Hsu, T.L. Chen, S.D. Lan, W.H. Lee and P.T. Wu, Appl. Phys. Lett. **56**, 2675 (1990).

45 K. Chen, Y.C. Chen, S.W. Lu, W.H. Lee and P.T. Wu, Physica **C 173**, 227 (1991).

46 K. Watanabe, N. Kobayashi, S. Awaji, G. Kido, S. Nimori, K. Kimora, K. Sawano and Y. Muto, Jpn. J. Appl. Phys. **30**, L1638 (1991).

47 K. Watabe, S. Awaji, N. Kobayashi, S. Nimori, G. Kido, K. Kimura and M. Hashimoto, Cryogenics **32**, 959 (1992).

48 K.H. Müller and C. Andrikis, Physical Review **B 49**, 1254 (1994).

49 J.L. Tholence, H. Noël, J.C. Levet, M. Potel, P. Gougeon, G. Chouteau and M. Guillot, Physica **C 153–155**, 1479 (1988).

50 M. Guillot, J.L. Tholence, O. Laborde, M. Potel, P. Gougeon, H. Noël and J.C. Levet, Physica **C 162–164**, 361 (1989).

51 A. Gerber, Z. Tranawski and J.J.M. Franse, Physica **C 209**, 147 (1993).

52 M.E. McHenry, H.S. Lessure, M.P. Maley, J.Y. Coulter, I. Tanaka and H. Kojima, Physica **C 190**, 403 (1992).

53 V.V. Chabanenko, A.I. D'yachenko, H. Szymczak and S. Piechota, Physica **C 273**, 127 (1996).

54 G.C. Han, K. Watanabe, S. Awaji, N. Kobayashi and K. Kimura, Physica **C 274**, 33 (1997).

55 J.L. Tholence, M. Saint-Paul, O. Laborde, P. Monceau, M. Guillot, H. Noël, J.C. Levet, M. Potel, J. Padiou and P. Gougeon, Studies of High Temperature Superconductors, Vol. 6, p. 37 (Nova Science Publishers, New York), Ed. A. Narlikar, 1990.

Chapter 6

Interactions Vortex-Vortex, Vortex-Defect and Vortex-Spin

1. Introduction

The pinning force per unit volume, F_p, is connected to the measurable critical current by the following equation: $F_p = J_c B$. However, this equation does not give information on the pinning mechanisms or on the way to increase the critical current value. It also does not join J_c to the pinning force, f_p, of an individual vortex. Let us consider as an example a homogeneous defects distribution in a superconducting material. These defects, which constitute real traps for the vortices, can be described in terms of potentials of interaction with the flux lines. They can attract or repulse the vortices, in other words, they can favor or oppose the electrons condensation in the Cooper pairs. The problem which arises is to know how these interactions between the vortices and the defects contribute to the determination of the pinning force density F_p. Right from the beginning, one is tempted to write:

$$F_p = N f_p \qquad (6.1)$$

where N is the number of these interactions. This is generally false because a rigid vortex lattice and a random distribution of defects lead in fact to no pinning phenomenon. Indeed, the sum of the individual pinning forces, of random orientations, is statistically zero. In other words, the interaction energy of an infinite surrounding is independent of the relative position of the rigid vortex lattice and the random defects present in this surrounding. But, the trapping of vortices exists in the superconductors and can be measured by different methods. The solution to this dilemma is that a process of pinning cannot take place except if the vortex lattice distorts itself. In this case, the total energy of the system is reduced by the distortion of the lattice, and the pinning occurs to prevent the growth of energy that would be provoked by the vortices movement. In the opposite limit or for a liquid vortex lattice, J_C is zero because in this case, it is necessary to prevent the movement of each vortex. The description of the rigidity of the vortices lattice is therefore essential to the understanding of the pinning phenomena which are the result of several types of interactions such as the vortex-defect interaction, the vortex-spin interaction and the vortex-vortex interaction. This last interaction can be studied by using the elastic constants of the vortices lattice.

2. Elasticity Theory of the Vortices Lattice

2.1. Isotopic Lattice

2.1.1. Local Elasticity

The concept of the elasticity of the flux lines was originally introduced by J.C. Maxwell in 1892[1] where the forces which act on a material in terms of compressibility of the magnetic field was described. As this idea is based only on the field energy density, it can be applied to the type II superconductors on the condition of replacing H by the law of the thermodynamic field $H(B)$.

The configuration of flux lines as well as their positions influences their elastic constants. These constants which characterize the interactions between the crystal and the lattice of vortices were at first described in terms of interaction between neighboring vortices by J. Silcox and R.W. Rollins[2], and by J. Friedel et al.[3] Finally it was R. Labush[4] who determined all the vortex lattice elastic constants at the equilibrium. In what follows, I will use the notation of Voigts for anisotropic crystals. In this notation, the constraints σ_i are connected to the distortions ε_i by:

$$\sigma_i = c_{ij}\varepsilon_i \tag{6.2}$$

Here, I take the values from 1 to 6 that correspond to the suffixes xx, yy, zz, yz, xz and xy, respectively. The tensor of the distortions ε_{ij} expresses the relative variations to the first order of the vortices position $\mathbf{r} = \mathbf{R} + \varepsilon$ in relation to the equilibrium position $\mathbf{R}$. These are the spatial derivatives of the displacement field ε, defined by:

$$\varepsilon_{ii} = \frac{\partial u_i}{\partial i} \qquad \text{and} \qquad \varepsilon_{ij} = \frac{\partial u_i}{\partial j} + \frac{\partial u_j}{\partial i} \tag{6.3}$$

The vortices are considered parallel to the z-axis and the symmetry of the vortices lattice is supposed hexagonal in the planes (x, y); its properties are isotropic. The invariance by translation according to the z direction reduces the general relations existing between the components of the constraints tensor and the components of the distortions tensor:

$$
\begin{pmatrix} \sigma_{xx} \\ \sigma_{yy} \\ \sigma_{yz} \\ \sigma_{xz} \\ \sigma_{xy} \end{pmatrix} = \begin{pmatrix} c_{11} & c_{12} & 0 & 0 & 0 \\ c_{12} & c_{11} & 0 & 0 & 0 \\ 0 & 0 & c_{44} & 0 & 0 \\ 0 & 0 & 0 & c_{44} & 0 \\ 0 & 0 & 0 & 0 & c_{66} \end{pmatrix} \begin{pmatrix} \varepsilon_{xx} \\ \varepsilon_{yy} \\ \varepsilon_{yz} \\ \varepsilon_{xz} \\ \varepsilon_{xy} \end{pmatrix}
\tag{6.4}
$$

with the restriction $2\,c_{66} = c_{11} - c_{12}$; thus, it remains only three independent elastic constants[5]. The uniform compression module, K, expresses the resistance to the compression in the plane (x,y) of the lines of flux aligned according to the **z**-axis (see Fig. 6.1). It is given by:

$$
K = \frac{c_{11} + c_{12}}{2}
\tag{6.5}
$$

$$
c_{11} \sim c_{12} \sim \frac{B^2}{4\pi}\frac{dH}{dB}
\tag{6.6}
$$

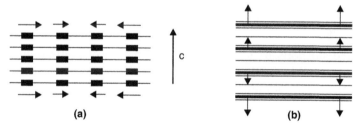

(a)   **(b)**

Figure 6.1. Examples of elastic distortions in an anisotropic superconductor. (a) Parallel compression to the superconducting planes. (b) Perpendicular compression to the superconducting planes. Contrary to what one could think, all the compressions are described by only one elastic coefficient even in presence of anisotropy.

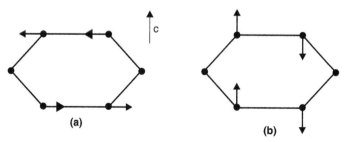

(a)   **(b)**

Figure 6.2. Examples of elastic distortions in an anisotropic superconductor. (a) Easy shearing. (b) Difficult shearing.

The tilting factor, c_{44}, determines the resistance to the tilting of vortices in relation to the applied magnetic field. It is given by:

$$c_{44} = \frac{B\,H(B)}{4\pi} \tag{6.7}$$

The shearing factor, c_{66}, expresses the resistance to the shearing of vortices in the planes (x,y) (see Fig. 6.2). This is the only constant that characterizes the symmetry of the lattice. It is connected to the energy difference between the square lattice and the triangular lattice and must vanish at H_{C2}. Close to the inferior critical field H_{C1}, the shearing factor is[4]:

$$c_{66} = \frac{1}{8\pi} \int_0^B B^2 \frac{d^2 H(B)}{dB^2} dB \tag{6.8}$$

For a type II superconductor ($\kappa \gg 1$) and for a value of the field H close to H_{C2}, c_{66} is given by[4]:

$$c_{66} \sim \frac{H_C^2}{8\pi\,\beta_A} \left(1 - \frac{H}{H_{C2}}\right)^2 \tag{6.9}$$

This expression shows that c_{66} approaches zero when H approaches H_{C2}: the lattice of vortices melts close to H_{C2}.

2.1.2. Non Local Elasticity

When the distortions are small in relation to a scale of length defined by an efficient penetration depth given by $\lambda' = \lambda/(1 - H/H_{C2})^{1/2}$, the elastic energy of the vortices lattice submitted to the constraints field ε_{ij} is given by the local expression:

$$W_{elastic} = \frac{1}{2} \int dV \left\{ c_{11} (\varepsilon_{xx} + \varepsilon_{yy})^2 \right. \tag{6.10}$$
$$\left. + c_{66} [(\varepsilon_{xy} + \varepsilon_{yz})^2 + (\varepsilon_{xx} - \varepsilon_{yy})^2] + c_{44} (\varepsilon_{xz} + \varepsilon_{yz})^2 \right\}$$

In the contrary case, E.H. Brandt[6] suggests to use the non-local elasticity whose the energy is expressed in the Fourier space in terms of displacement field $\upsilon(\mathbf{k})$:

$$W_{elastic} = \frac{1}{2} \int_{Z.B.} \frac{d^3k}{8\pi^3} \sum_{\alpha,\beta} \upsilon_\alpha(\mathbf{k})\, \upsilon_\beta^*(\mathbf{k})\, \Phi_{\alpha\beta}(\mathbf{k}) \tag{6.11}$$

where α and β describes the Cartesian coordinates and $\Phi_{\alpha\beta}(\mathbf{k})$ represents the elastic matrix. The integration is done on the first zone of Brillouin

(Z.B.) for $-\infty < k_z < \infty$. $\Phi_{\alpha\beta}(\mathbf{k})$ and $\upsilon(\mathbf{k})$ are periodic quantities. In the case where $k \ll k_{Z.B.}$ with $k_{Z.B.} = 2H/H_{C2}\xi$, $\Phi_{\alpha\beta}(\mathbf{k})$ is given by:

$$\Phi_{\alpha\beta}(k) = k_\alpha \, k_\beta \, [c_{11}(k) - c_{66}]$$
$$+ \delta_{\alpha\beta} \, [(k_x^2 + k_y^2) \, c_{66} + k_z^2 \, c_{44}(k)] \tag{6.12}$$

From where one deducts:

$$c_{11}(\mathbf{k}) - c_{66} = \frac{H_{C2}^2}{4\pi} \times \frac{\left(\dfrac{H}{H_{C2}}\right)^2}{1 + \lambda'^2 \, k^2} \times \frac{1 - \dfrac{1}{2\kappa^2}}{1 + \dfrac{\lambda^2 \, k^2}{2\kappa^2}} \tag{6.13}$$

and:

$$c_{44}(k) = \frac{H_{C2}^2}{4\kappa} \left[\frac{\left(\dfrac{H}{H_{C2}}\right)^2}{1 + \lambda^2 \, k^2} + \frac{\dfrac{H}{H_{C2}}\left(1 - \dfrac{H}{H_{C2}}\right)}{\kappa^2} \right] \tag{6.14}$$

where $k^2 = k_x^2 + k_y^2 + k_z^2$. Expressions 6.13 and 6.14 are valid for $\kappa^2 H/H_{C2} \gg 1$ or for $H/H_{C2} \gg 0.3$. In the two cases, the magnetic fields of the vortices overlap. These expressions also show that c_{11} and c_{44} are dispersive; they depend on the length of wave $2\pi/k$ of the field of constraints. It is useful to underline at this level that the energies put in play for the shearing are generally lower than the compression energies of the vortex lattice. Therefore, if the vortices lattice has the choice between the one and the other of the two possible distortions in the (x,y) plane, it will opt for the shearing distortions rather than for the compressions that would cost many elastic energy. As c_{66} is very little dispersive, it is not then necessary to call on the non-locality to calculate the elastic energies of shearing. In the non-local elasticity, the displacement of vortices due to the pinning sites is more important than the one calculated in the framework of the local elasticity, giving a more elevated value of the calculated critical current density. Besides, the coupling and re-pasting processes of vortices are encouraged and the effect of fluctuations is more important[7].

2.2. Anisotropic Lattice

The extension of the non-local elasticity theory to an anisotropic lattice of vortices is laborious enough because if the elastic energy is always described by Equation 6.11, the elastic matrix (see Equation 6.12) in this case takes a complex form (except for $B//c$). Indeed, the elastic energy of a distortion will depend on the vortices direction in relation to a reference direction. Besides, the three constants c_{11}, c_{44} and c_{66} are not sufficient to describe the elastic properties of such a lattice. Thus, an undulation of the vortices parallel to the superconducting planes will cost, for example, less energy than a perpendicular undulation to these last from where the necessity to introduce two undulation constants c^{ab}_{44} and c^{c}_{44} (see Fig. 6.3). This is similarly for the shearing modes for which one defines two shearing constants corresponding to two different symmetry distortions; one easy and one difficult (see Fig. 6.2). Thus, one may add new elastic constants which describe mixed modes where the distortions of shearing or of compression are coupled to distortions of undulation[8-9].

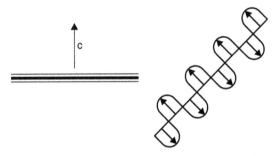

Figure 6.3. An undulation distortion in an anisotropic superconductor where the field of small displacements has a component parallel to the *c*-axis. This undulation is more difficult than the one where the field of small displacements has components according to *x* and *y* (such a field would be perpendicular to the plane of the figure).

3. Different Approaches of the Vortices Trapping

When one applies a magnetic field, the presence of defects in a type II superconductor leads to the occurrence of a flux gradient that puts the vortices lattice under tension (for example, in an increasing field, these defects create an important vortices density close to the sample surface than in its center, see Chapter 3, Paragraph 5). The expression of the magnetic force per unit volume F_V is given in the approximation of the local elasticity by[10]:

$$F^i_V = \sum_j \frac{\partial \sigma_{ij}}{\partial x_j} = \sum_{j,k,l} c_{ijkl} \frac{\partial \varepsilon_{kl}}{\partial x_j} \tag{6.15}$$

where σ_{ij} represents the tensor of the constraints, ε_{kl} the tensor of the distortions and c_{ijkl} the tensor of the elastic constants of the vortices lattice. If there is no curvature of the vortices, Equation 6.15 cuts down to:

$$F^i_V = K \frac{\partial}{\partial x_i} \left(\sum_m \varepsilon_{mm} \right) \sim K \frac{\partial B}{B \partial x_i} \tag{6.16}$$

where $K = (c_{11} + c_{12})/2$, it represents the factor of compression of the straight vortices lattice. By replacing the coefficients c_{11} and c_{12} by their values

(Equation 6.5), one gets the relation of J. Fridel[11]:

$$F_V^i = \frac{B}{4\pi} \frac{\partial H}{\partial B} \frac{\partial B}{\partial x_i} \qquad \text{where} \qquad i = x \text{ et } y \qquad (6.17)$$

When this magnetic force becomes higher than the vortices pinning force F_p, these latter move in order to reduce the gradient of B until equilibrium gets settled again between these two forces, this is the concept of the critical state which was studied in Chapter 5. The first attempts aiming to calculate the vortices pinning force leave from the local displacement of a vortex of a distance u_0 in relation to its equilibrium position under the action of an elementary force of interaction, the elasticity of the lattice resulting in a force of restoring. In 1968, H.T. Coffey[12] elaborated a simple model which equals the pinning force to the restoring force exercised by the vortices lattice. With an empiric value of u_0, he showed that the sum of the pinning forces $F_p = J_C\ B$ is proportional to the density of surface, n_p, of the pinning centers:

$$F_P = \frac{n_p\, H_C}{4\pi\, \lambda} \left(\frac{H}{H_{C2}} \right)^{1/2} \left(1 - \frac{H}{H_{C2}} \right) \qquad (6.18)$$

This expression described the experimental results obtained on Nb_3Sn and NbTi on a large field interval. In 1969, W.A. Frietz and W.W. Webb[13] noted that in presence of pinning centers distributed in a random way in a superconductor, the vortices lattice must distort itself in order to fit to these defects and minimize its energy. This energy depends on the position of the vortices in relation to the distribution of the traps and a pinning force that must be function of the faculty of the lattice to distort itself in order to deviate from its equilibrium position under the action of the pinning force f_c. By considering the gap $u_0 \sim f_c/C$ where C is the rigidity of the vortex and by supposing that the number of centers N interacting with the vortex is proportional to u_0, they arrive to:

$$F_P = Nf_C \sim \frac{f_C^2}{C} \qquad (6.19)$$

where C is a function of B. This equation reveals a quadratic dependence of F_p with f_c: the vortices can be trapped as well by attractive forces as by repulsive forces. R. Labush[14] proposed, in 1969, a statistical theory in order to calculate the total force F_p starting from the force, f_c, of the elementary interaction vortex-defect. He considered the vortices lattice like a surrounding characterized by three elastic constants c_{11}, c_{44} and c_{66}. By looking for the solutions of the general equation of forces acting on a vortex that lead to an unstable equilibrium in a random distribution of the trapping centers of which the interaction range vortex-defect is lower than the distance between vortices, he succeeds to the expression of the following pinning force per unit volume:

$$F_p = \frac{n_p f_C^2}{8\sqrt{\pi}} \left(\frac{B}{\Phi_0} \right)^{3/2} \left(\frac{1}{\sqrt{c_{11} c_{44}}} - \frac{1}{\sqrt{c_{66} c_{44}}} \right) \tag{6.20}$$

This expression is used as theoretical support to the different laws of scale which are generally used to interpret the variation of the critical current density J_C as a function of the field in the case of the strong pinning mechanism:

$$J_C(B) = J_{C0} \left(\frac{H}{H_{C2}} \right)^p \left(1 - \frac{H}{H_{C2}} \right)^q \tag{6.21}$$

where p and q designate characteristic exponents. For recall, in the strong pinning mechanism, the pinning forces are stronger than the interactions between the vortices; the lattice of vortices distorts itself in optimal way occupying all the available trapping sites (see Fig. 6.4a).

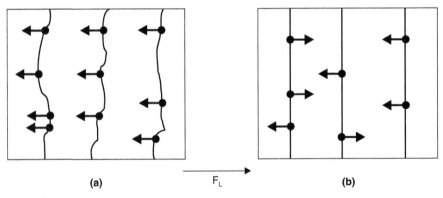

(a) F_L (b)

Figure 6.4. Pinning of the vortices lattice in presence of a macroscopic Lorentz force. (a) Case where the pinning forces predominate the interactions between vortices. (b) Case of a perfectly rigid vortices lattice.

The microscopic pinning forces act in a coherent way, and the obtained macroscopic pinning force is the sum of all pinning forces exerted by all the defects contained in the material. This is precisely the common denominator of the theoretical approaches of pinning mentioned above[5,15]. In the opposite limit where one has to deal with a perfectly rigid vortices lattice, the consequence of the pinning forces is statistically equal to zero (see Fig. 6.4b). In reality, this is the intermediate case that presents itself where the pinning forces are comparable to the elastic forces. The macroscopic pinning force, which determines the value of the critical current density, is then the result of a compromise between the pinning and the interactions between the vortices. This case was studied by using the collective pinning theory of A.I. Larkin and Yu. N. Ovchinnikov[16]. They supposed that the defects present in the material destroy the long-distance order of the vortices lattice, but that the short-range order remains intact in correlated regions of volume $V_C = R_C^2 L_C$ where L_C and R_C represent the lengths of correlation parallel and perpendicular to the vortices, respectively. In one of these areas, the $n_p V_C$ defects when randomly distributed exert a force on V_C whose statistical fluctuation gives an amplitude equal to $(n_p V_C f_p^2)^{1/2}$, where n_p is the average density of the defects in the sample and where f_p^2 ($= \langle f_i^2 \rangle$) represents the average square of forces f_i exerted on the vortices by each of the individual defects indexed by i, and of which the sum of all the forces per unit volume is:

$$F_P = J_C \, B = \frac{\langle n_p \, V_C \, f_p^2 \rangle^{1/2}}{V_C} = \left(\frac{W}{V_C} \right)^{1/2} \tag{6.22}$$

where $W = n_p \langle f_i^2 \rangle$, it represents the mean square of the fluctuations per unit volume. The volume V_C is the region in which the squared root of the correlation function $g(\mathbf{r})$ of the field of little displacements $u(\mathbf{r})$ is weak at the range r_p of the trapping forces, of the order of max(ξ, R_d) for $H \ll H_{C2}$, where R_d is the average radius of the defects, thus:

$$g(\mathbf{r}) = \langle [\mathbf{u}(\mathbf{r}) - \mathbf{u}(0)]^2 \rangle \leq r_p^2 \tag{6.23}$$

This condition gives for the three-dimensional volume V_C an ellipsoid of volume $(4\pi/3) R_C^2 L_C$ and for the two-dimensional volume V_C a circle of volume πR_C^2, or more precisely in the thin layers, the volume V_C is a cylinder whose height is equal to the thickness of the layer. This gives an idea of the effect of the dimensionality of the system on the critical current, an effect which becomes very important for the high-T_C oxides. This theory also shows that the average pinning force F_p depends on the mean square of the elementary pinning forces, which means that the bundles of vortices can be trapped by attractive forces as well as by repulsive forces and that there is no threshold for the elementary forces in the meaning of $F_p = 0$ for $f \leq$ threshold. The most important success of this theory is the interpretation of the increase of $J_C(H)$ in the type-II superconductors when the applied field has a value less than H_{C2}. This collective pinning concept is very useful for other problems such as the pinning of the dislocations by atoms[6] or the pinning of the Bloch walls inside the magnetic materials[17]. It is to be noted that this theoretical problem of summation of randomly distributed weak pinning forces was independently studied by P.A. Lee and T.M. Rice[18] who considered the pinning of the waves of density of loads. The strong pinning mechanism (a defect per correlation volume) evidently cannot be studied by this theory. It is also possible to approach the phenomenon of the vortices pinning through the free energy of Ginzburg-Landau. Indeed, the defects present inside the superconductor, locally modify the physical properties of the material and consequently affect the modulation of the parameter of order or the microscopic magnetic field. This allows distinguishing between two kinds of elementary mechanisms of the vortex coupling in the material, the magnetic interactions and the core interactions. The magnetic interactions generally originate an interface superconductor/normal metal, parallel to the applied field. This interface, which can be the surface of the sample or the surface of a precipitate of a size greater than or equal to the penetration depth, modifies the distribution of the current which circulates around the core of the vortex so that the normal component on the surface vanishes. This problem is reduced to the attraction of the vortex by its anti-vortex that is to say its image in relation to the interface. This interaction is opposed to the repulsion of the currents which shield the external magnetic field. Futher, this interaction mechanism is weak for the type-II superconductors; it generally disappears for the high magnetic fields. The interactions of the core directly act on the spatial variation of the parameter of order. Around the defect (punctual, linear or planar), the physical properties of the superconductor like the density, the elasticity, the electron-phonons coupling or the electronic mean free path, are modified. A modification of the first three properties leads to a local variation of the critical temperature T_C whereas a variation of the mean free path mainly acts on the parameter of Ginzburg-Landau κ. This distinction gives completely different dependences in field and in temperature for the pinning forces. It is then possible to obtain an expression of the vortex-defect interaction energy starting from the function of Ginzburg-Landau by considering that the defects affect the coefficients α, β as well $\mathbf{A}$ and Ψ. It is given to the first order by[5]:

$$\delta E = \int \frac{H_C^2}{8\pi} \left(-\frac{\delta H_{C2}^2}{H_{C2}^2} |\psi|^2 + \frac{1}{2} \frac{\delta \kappa^2}{\kappa^2} |\psi|^4 \right) dV \qquad (6.24)$$

This expression was obtained by taking into account only the modifications $\delta\alpha$ and $\delta\beta$ of the coefficients α and β. In addition, it is in fact valid only if the size of the defect is lower than the coherence length ξ. δE represents the energy of the vortex-defect interaction at the distance of variation of the upper critical field δH_{c2} and of the Ginzburg-Landau parameter $\delta\kappa$. The maximal gradient δE gives the elementary pinning force f_c. Before closing this paragraph, it is to be noted that among the phenomena of the vortices pinning, the plasticity of the vortices lattice whose the role is crucial for the high-T_C superconductors may be quoted because the relatively weak values of their elastic coefficients favor the formation of the dislocations either by thermal fluctuations either by internal constraints created by the pinning and the force of Lorentz.[19]

4. Critical Currents

4.1. Intra-granular Currents

The intra-granular currents are usually deducted by using magnetic measurements. These currents are generally of the order of 10^6 to $10^7 A/cm^2$, at low temperatures and fields. These current values were firstly observed, in 1987, by M. Ousséna et al.[20] in the $La_{1.85}Sr_{0.15}CuO_4$ compound before the discovery during the same year of the $YBa_2Cu_3O_{7-\delta}$ compound by C.U. Chu et al.[21], and subsequently confirmed in the other high-T_C oxides, especially by transport measurements in the thin layers of Y(123) by P. Chaudary et al.[22]. The critical currents deducted from the magnetic measurements sharply fall according to the temperature as well as according to the field at high temperatures. In the case of $YBa_2Cu_3O_{7-\delta}$ for example, the critical current decreases by two orders of magnitude when the temperature goes from 4.2 to 77K[23–30]. This situation is generally much more dramatic for the bismuth[31–34] and thallium[35–36] compounds.

4.2. Inter-granular Currents

Contrary to the intra-granular currents (i.e. in the single crystals and in the isolated grains), the inter-granular currents are mainly obtained by transport measurements. There are two reasons for this. The first reason is that in general the inter-granular current densities are rather low and therefore relatively easy to measure directly because the Joule heating is moderate. The second reason is that owing to the complexity of the problems of weak links between the grains, there are currently no quantitative models that connect the hysteresis loop of the granular materials to the macroscopic critical current. This is also true for the susceptibility measurements.

4.3. Influence of Defects on the Critical Current

The classical method to improve the critical current is to add to the superconducting matrix an appropriate number of trapping centers, of dimensions comparable to the coherence length ξ. The concentration of these trapping centers must be sufficiently high to pin a maximum of flux lines without significantly disturb the critical temperature and other basic parameters such as ξ and λ. In the conventional superconductors, this is often carried out by the addition of ordinary metals and the subsequent application of the appropriate heat and mechanical treatments. Unfortunately, this situation becomes complicated for the high-T_C

superconductors because the superconducting properties are very sensitive to the slightest deviation from the stoichiometry of the oxygen and copper concentration. Indeed, very weak variations can increase the critical current density. However, there exists a threshold beyond which these variations deteriorate this critical current. The value of this threshold is not an intrinsic quantity but depends on the local distribution of defects and therefore on the heat treatment and the other factors which affect this distribution because the critical current density strongly depends on the chemical nature of the defects as well as on their shapes and their dimensions. It seems that the most favorable dimension for the critical currents is of the order of 2 to 6 times the coherence length. The defects of this size allow to reduce the electromagnetic energy of the vortices without significantly affecting the superconducting state[37]. This is, in fact, a very important point which requires more experimental and theoretical researches.

4.4. Butterfly Effect

One of the most surprising features of the $M(H)$ cycles of the superconductors is their butterfly shape. This particular shape was observed in the conventional superconductors[38-39] as well as in the high-T_C superconductors[40] by magnetization[41], transport[42] and torque[43] measurements. Many explanations of this effect were proposed in the literature. It is not a result of the demagnetizing effects. Some authors attribute this effect to the trapping effects. Further studies are needed to fully understand its physical origins.

4.5. External Surface-Vortex Interaction

The vortices pinning by the external surface of the sample was extensively studied in the conventional superconductors. Its consequences on the magnetic properties of the conventional materials were studied by C.P. Bean and J.D. Livingston[44] and analyzed in detail by H. Ulmaeir,[45] in particular in connection with the hysteresis loops of the type II superconductors which result in what is known as the minor cycles. From a theoretical point of view, the interaction of the sample surface with the vortices can be established mainly in three ways: (i) the thermodynamic or intrinsic surface barriers (ii) the surface imperfections (iii) the extrinsic pinning of the trapping centers near the surface.

4.6. Perfect Surface-Vortex Interaction

A perfect surface leads to a significant distortion of the fields and currents of the flux lines located inside the sample, over a distance of the order of λ or less, calculated starting from the surface. The physical requirement that the vortex currents must be necessarily tangential to the surface leads to the image of the vortex of opposite polarity to the real vortex (anti-vortex). These two vortices interact in the same way as those two real vortices except here, the interaction is attractive whereas it is usually repulsive in a vortices lattice.

5. Lock-in Transition

The lock-in transition of the vortices is a specific phenomenon to the layered superconductors. It occurs for magnetic fields whose direction is very close to the superconducting planes. The vortices then prefer to align along these planes because the gain in energy of core becomes higher than the cost in elastic energy. Experimentally, the lock-in transition results in an anomaly of the reversible torque: when the direction of the vortices switches to become parallel to the superconducting planes, the angle between the field direction and the vortices increases leading to an increase of the torque[46]. This anomaly was observed by us where we were able to highlight this transition in the quasi-2D system as well as in the quasi-3D system[47] (see Fig. 6.5). Other measurements permit to highlight the lock-in transition such as alternative susceptibility measurements[48] and transport measurements. In particular, the sharp drop of the resistivity observed by W.K. Kwok et al.[49] was interpreted by these authors like coming from the disappearance of the stairs during the locking of vortices near the direction of the superconducting planes.

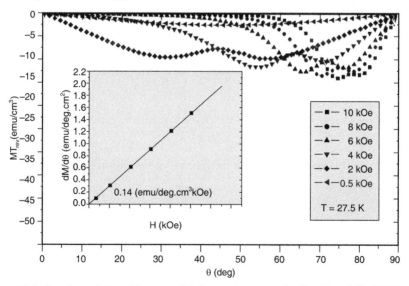

Figure 6.5. Angular variation of the reversible transverse magnetization M_{Trev} at $T = 27.5$K and $H = 0.5, 2, 4, 6, 8, 10$ kOe. The inset shows the slopes of M_{Trev} near $\theta = 90°$ for several values of the applied field. The straight line is a fit with $dM_{xrev}/d\theta = AH$ with $A = 0.14$ emu/cm³/deg/kOe (after S. Khene, 2004 [47]).

6. Pinning Theory of Feigel'man, Geshkenbein and Larkin

Let us recall that in the London model, it is possible to calculate an anisotropic physical quantity Q from its expression in the isotropic case Q by using a scaling law developed by G. Blatter et al.:[50]

$$Q(\theta, H, T, \xi, \lambda, \gamma, S) = S_Q \, Q(\varepsilon(\theta)H, \gamma T, \xi, \lambda, \gamma \, S) \tag{6.25}$$

where $\theta, \xi, \lambda, \gamma, H, T, S$ represent the angle between the applied field and the c-axis of the crystal, the coherence length, the penetration depth of London, the anisotropy factor, the field, the temperature and the action, respectively. $S_Q = \gamma$ for the volumes, the energies, the temperature and the action. For the magnetic field $S_H = 1/\varepsilon(\theta$ avec $\varepsilon(\theta = (\sin^2\theta + \gamma^2 \cos^2\theta)^{1/2}$. Although limited to the three-dimensional collective pinning, this transformation permitted to calculate with a great success the angular dependence of the elastic coefficients of the vortices lattice[51]. In the case of the vortices pinning of Josephson (i.e., two-dimensional), this scaling transformation is no longer valid because within this system several types of pinning exist. Thus, in the region of low fields and low temperatures, M.V. Feigel'man et al.[52] considered the weak pinning of the quasi-two-dimensional vortices in the direction of the c-axis and showed the existence of a 2D pinning of isolated vortices. The critical current is independent of the field and close to its maximum value:

$$J_C^{s.v.} = \frac{c \, U_p}{d \, \Phi_0 \, \xi} \tag{6.26}$$

where U_p is the depth of the potential well and d the distance between the superconducting planes. This relation is valid only for $U_p \gg U_{int.}$ $U_{int.}$ is the interaction energy given by:

$$U_{int.} = \frac{\Phi_0^2 \, S \, \xi^2}{16 \, \pi^2 \, \lambda^2 \, a_0^2} \tag{6.27}$$

a_0 represents the average distance between two neighboring vortices:

$$a_0 = \sqrt{\frac{\Phi_0}{B}} \tag{6.28}$$

Beyond a critical field, H_1 given by:

$$H_1 = \frac{U_p \, \Phi_0}{U_{int.} \, a_0^2} \tag{6.29}$$

the pinning is still two-dimensional but becomes time collective. In each layer, the vortex bundles of the correlation length:

$$R_C = \frac{U_{int.}}{U_p} a_0 \tag{6.30}$$

are independently trapped of each other. The current is proportional to the bundles size:

$$J_C^{2D,\, col.} \sim J_C^{s.v.} \frac{a_0}{R_C} \tag{6.31}$$

When the bundles radius becomes comparable to the Josephson length $\Lambda = \gamma d$, the trapping regime becomes three-dimensional. The transitions between two-dimensional and three-dimensional pinning regimes are defined by the condition:

$$U_p \geq U_{int.} \frac{a_0^2}{\Lambda^2} \tag{6.32}$$

The critical current in the three-dimensional regime is given by:

$$J_C^{3D} = J_C \left(1 + \frac{k_B T}{U_{int.}}\right)^{-5/2} =$$
$$J_C \left(1 + k_B T \frac{16\, \pi^2\, \lambda^2}{\Phi_0\, d\, B}\right)^{-5/2} \tag{6.33}$$

If the coupling between the superconducting planes becomes weak, the vortices lattice melts under the action of the thermal fluctuations. Above the melting temperature, the vortices are no longer pinned, static magnetization becomes zero. At low temperatures, the vortices glass regime appears where each vortex is on average immobile but the lattice loses its coherence at long distance. When the temperature increases to become comparable to the energy of activation, one observes the thermally activated flux flow regime. The thermal activation causes a rapid decrease of the critical current with the temperature. The Anderson model[53] expects a variation of J_c as a function of the temperature of the type:

$$J_c \sim J_c(0) \left(1 - \frac{T}{T_c}\right) \tag{6.34}$$

The theory of the collective pining expects a law of the type[54]:

$$J_C = J_C^* \left[\frac{U^*}{T \, ln\,(t/t_0)} \right]^\alpha \tag{6.35}$$

where J^* and U^* are constants independent of the temperature. α is equal to 7, 2/3 and 9/7 for the trapping of an isolated vortex, a collective pinning and for $J_C \ll J_C (T = 0K)$, respectively. The critical current vanishes for a temperature T_i below the critical temperature T_C. T_i characterizes the irreversibility line[55]. For the bismuth compound Bi(2212), this line is located at $T < T_C/2$ whereas for the Y(123) compound, it is observed near T_C[56]. It also depends on the experimental conditions.

7. Superconductivity and Ferromagnetism Coexistence in YBa$_2$Cu$_3$O$_{7-\delta}$ Nanoparticles

7.1. Introduction

Nanoparticles study is a multidisciplinary science which gathers together chemists, physicists, biologists, physicians and engineers. It stimulates the interest of scientists and industrialists, because it opens up new perspectives in the understanding and development of new materials whose properties are intimately related with the surface-volume ratio of the particle. Indeed, at this scale, the surface properties are predominant. Superconducting nanoparticles, characterized by a size comparable to the penetration depth λ and/or the coherence length ξ, are very important systems, in both basic scientific research as well as in advanced nanotechnology. Indeed, they exhibit many novel physical properties that differ significantly from those of corresponding bulk materials. Ferromagnetism is one appealing property observed in nanoparticles of non-magnetic metal oxides, non-magnetic metal nitrides and chalcogenides, metallic materials and in diluted magnetic semiconductors[57]. It was claimed that superconductivity and ferromagnetism were incompatible until the establishment of their coexistence in UGe$_2$[58]. Ferromagnetism in both superconducting materials and non-magnetic materials is believed to be due to intrinsic defects such as oxygen vacancies in the material[59-60]. Shipra and co-workers[61] and Sundaresan et al.[57,62-63] reported room temperature ferromagnetism in YBa$_2$Cu$_3$O$_{7-\delta}$ nanoparticles with a possible existence of the ferromagnetism at $T = 90K$ ($T_C = 91K$). Hasanain et al.[64] investigated the particle size dependence of the magnetic and superconducting properties of nanoparticles of YBa$_2$Cu$_3$O$_{7-\delta}$. They explored the relationship between the sizes of the nanoparticles and the presence of magnetic and superconducting effects. They also studied the effect of the ferromagnetic phase on the superconducting properties and discussed a possible coexistence of the surface ferromagnetism and the bulk superconductivity at lower temperatures. Zhu et al.[65] showed that the ferromagnetism in nanoparticles of $YBa_2Cu_3O_{7-\delta}$ is associated with the surface oxygen vacancies. They also confirmed the presence of ferromagnetism at room temperature and demonstrated the coexistence of ferromagnetism and superconductivity in YBCO nanoparticles at 5K. The present experimental study considers superconductivity and ferromagnetism coexistence in nano-sized powders of $YBa_2Cu_3O_{7-\delta}$ at 4.2K and aims at establishing whether superconductivity and ferromagnetism uniformly coexist or are phase-separated in nano-domains[66].

7.2. Crystallographic Properties

7.2.1. Yttrium Iron Garnet ($Y_3Fe_5O_{12}$)

The ferrites garnets of rare earths were discovered by Bertaut and Forrat at Grenoble (France)[67] and by Geller and Gilleo at Murray Hill (USA)[68]. The « garnet » term designates compounds of the crystallographic structure similar to the well-known precious stone, the orthosilicate $\{Ca_3\}$ $[Al_2](Si_3)O_{12}$. The first deepened study of this structure was made by Yoder and Keith[69]. These compounds form an important group of ferromagnetic insulating materials of general formula $\{R_3^{3+}\}[Fe_2^{3+}](Fe_3^{3+})O_{12}$ called RIG (Rare-earth Iron Garnet) where R denotes a rare-earth element or yttrium. The space group of the body-centered cubic unit cell is $Ia\bar{3}-(O_h^{10})$ $N°230$[70], it contains 160 ions. A detailed description of their properties can be found in reference 71. Because of their particular properties, the ferromagnetic garnets are dealt with in many theoretical and experimental researches (Difference between the experimental values of the magnetic moment of the iron Fe^{3+} deduced from magnetic measurements and neutrons diffraction experiments[72], applications in electronic devices, such as circulators and phase shifters for microwave and magneto-optical devices[73]). Besides, YIG has a well-regulated saturation magnetization, a low dielectric loss tangent ($\tan \delta$) in microwave regions and a small width (ΔH) in the ferromagnetic resonance[74]. The ions distribution on different crystallographic sites is fixed. It permits them to have near perfect compounds. Only yttric earth can give pure garnets. The unit-cell parameter is about 12Å and weakly decreases in the direction of lanthanides. It is thus possible to obtain a set of compounds which have crystalline properties very neighbors (same space group, with somewhat different parameters) and very different magnetic properties (where the interest for the applications to have, by substitutions, compounds and whose magnetic properties can be adjusted).

7.2.2. Ferrimagnetic Spinels

The general chemical formula of the ferrites with a spinel structure is: X^{2+} $Y_2^{3+} O_4^{2-}$, where X is a bivalent cation (Ni^{2+}, Fe^{2+}, Mn^{2+}) and Y is a trivalent cation (Fe^{3+}, Al^{3+}). The main ferrimagnetic substances of spinel structure are: the magnetite, Fe_3O_4 (the oldest magnetic materials), the nickel ferrite, $NiFe_2O_4$ (material of reference for fundamental studies) and manganese-zinc ferrite, $MnZnFe_2O_4$ (known for its very high value of permeability). The crystallographic structure of these ferrites reproduces one of the mineral spinel: the magnesium aluminum mixed oxide $MgAl_2O_4$. It is determined (as in all oxides) by the arrangement of oxygen ions. These ions form a face-centered cubic lattice whose interstitial sites are partially occupied by metallic ions. The unit-cell contains eight molecules and defines 64 tetrahedral sites, surrounded of 4 oxygen ions of which only 8 are occupied by metallic ions, and 32 octahedral sites, surrounded of 6 ions oxygen of which only 16 are occupied by metallic ions (see Fig. 6.6).

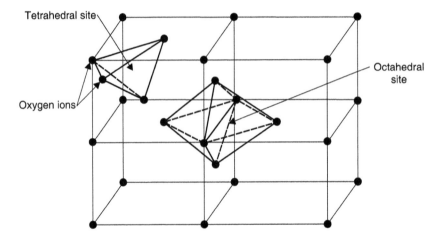

Figure 6.6. Tetrahedral and octahedral sites.

The unit-cell with spinel structure has a cubic symmetry. In the case of usual spinels, the bivalent ions occupy the tetrahedral sites and the trivalent ions occupy the octahedral sites. However, in the so-called "inverse" spinels, the bivalent ions occupy the octahedral sites and the trivalent ions distribute themselves between the octahedral sites and the tetrahedral sites.

7.2.3. Ferrimagnetic Garnets

The crystalline structure of the ferrimagnetic garnets is cubic, but more complex than one of the spinels. The general formula of compounds with a garnet structure is: $X_3^{2+} Y_2^{3+} Z_3^{4+} O_{12}^{2-}$, where X is a bivalent cation (Ca^{2+}, Mn^{2+}, Fe^{2+}), Y a trivalent cation (Al^{3+}, Cr^{3+}, Fe^{3+}) and Z a tetravalent cation (Si^{4+}). Its crystallographic structure reproduces one of the natural garnets in the chemical composition: $Ca_3Al_2Si_3O_{12}$. The most well known of the ferrimagnetic garnets is the yttrium iron garnet commonly called YIG which results from a synthesis in a laboratory, by replacing the bivalent and tetravalent ions by the trivalent ions: $Y_3^{3+} Fe_2^{3+} Fe_3^{3+} O_{12}^{2-}$. In the garnet structure, the metallic ions distribute themselves in three sorts of interstitial sites: tetrahedral sites, octahedral sites and dodecahedral sites (surrounded by 8 oxygen ions).

However, contrary to the spinels, all tetrahedral and octahedral sites are occupied by metallic ions thus procuring a greater stability to the garnet structure. In the case of YIG, the dodecahedral sites are occupied by the ions of yttrium which have a big atomic radius.

7.3. *Experimental Techniques*

To carry out our measurements, we used four techniques described as follows:

7.3.1. *Conventional Magnetometer*

The magnetization measurements as a function of the field $M(H)$ and the temperature $M(T)$ were carried out by means of a conventional magnetometer. Its main features are:[75-77]
- Accuracy: 5×10^{-7}A.m² (5×10^{-5} emu)
- Range of applied magnetic fields: ± 10.8T
- Range of temperatures: 1.5K to 300K
- Stability of the temperature: 0.01K
- Precision on the temperature: $dT = (0.2 \pm 0.002)$T (in Kelvin)

This machine is fully automatic. It is composed of a cryostat made of aluminum, which is thermally isolated (see Fig. 6.7). The numeric regulation of the temperature by means of a thermometer in carbon, supplied by a constant voltage using the four wires method; it permits for control and to measure the temperature in the range 1.5K – 300K with a stability of 0.01K. The thermal variation of this system uses helium.

7.3.2. *X-rays Diffractometer*

The diffractometer of x-rays is used in nearly all the domains of x-ray diffractometry such as the analysis of powders and phases, the study of structures and the measurements of constraints and textures. It is controlled by computer and is equipped with distinct moving parts which can be coupled in an electronic manner or can be driven independently.

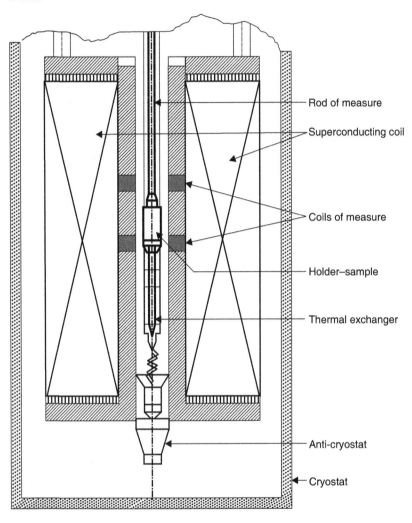

Rod of measure

Superconducting coil

Coils of measure

Holder–sample

Thermal exchanger

Anti-cryostat

Cryostat

Figure 6.7. Schematic representation of the magnetometer

The radiation emitted by the linear home of the x-ray tube is diffracted by the sample and is captured by the detector. The sample undergoes a rotation with a constant angular speed and the detector turns with a double angular speed around the sample. In this manner, the diffraction angle (2θ) is always equal to double of the incident angle (θ). Whenever the Bragg condition is satisfied, the primary radiation is reflected by the sample to the detector. With the help of the detector and the electronic measure which are connected, the intensity of the reflected radiation is measured, while the angular position of the reflected radiation is displayed by the order of the goniometer. In this manner, one gets impulses rates or a diffraction diagram. In order to assure a great clarity to the diagrams, the K_β stripes can be attenuated with the help of a filter or a secondary monochrometer.

7.3.3. Atomic Force Microscope (AFM)

The atomic force microscope (AFM) explores the strengths of the surface at the nanometric scale owing to the weak size of its probe (< 10nm). This device produces the image of an electrical, insulating or conducting surface.

A. Involved Forces

i)   Van der Waals Strength

Whatever the operated conditions; two forces are always present in the tip-surface interaction, the Van der Waals force and the chemical force. The Van der Waals strength is an attractive strength which acts over a long distance (see Fig. 6.8). It is due to the fluctuations of the electrical dipole moment between the atoms constituting of the probe and surface. The resulting potential U_{vdW} can be written as follows:

$$U_{VdW} = -\frac{A}{d^6} \qquad (6.36)$$

where A is a constant and d the distance between two considered atoms. In a sphere, $A = HR/6$ where H is the constant of Hamaker and R the sphere ray.

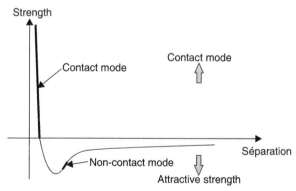

Figure 6.8. Van der Waals force as a function of the tip-sample distance. The microscope can be operated in two modes shown in the figure by bold lines.

ii)   Chemical Strength

The short-range chemical strength is repulsive ($d < 1nm$). This force is a consequence of the principle of exclusion of Pauli which prohibits the interpenetration of electronic clouds of atoms of the tip and the surface. By taking into account these two strengths, for a distance tip-sample smaller than the inter-atomic distance (0.5nm), the potential of interaction is described by the potential of Lennard-Jones U_{LJ} (see Fig. 6.9):

$$U_{LJ} = -\varepsilon \left[2\left(\frac{\sigma}{d}\right)^6 - \left(\frac{\sigma}{d}\right)^{12} \right] \tag{6.37}$$

where ε and σ are the energy and the distance of the chemical link of equilibrium. This potential reveals two separate regimes, an attractive régime with a variation in d^{-6} due to Van der Waals strengths and a repulsive régime with a variation in d^{-12}.

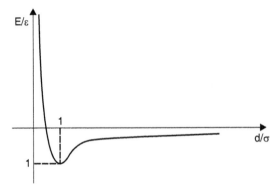

Figure 6.9. Representation of the empiric potential of Lennard-Jones corresponding to the potential between two non-loaded atoms. The minimum has for coordinates the inter-atomic distance σ and the energy ε of the atomic link.

iii) Electrostatic Strength

The electrostatic strength is present when the probe and the sample are conductors and present a difference of potential V. This strength is attractive; it acts over a long distance. It is dependent on the sample geometry and the probe. The electrostatic potential $U_{elec.}$ is written as a function of the capacity $C(d)$ which depends on the distance holder-sample d as follows:

$$U_{elec.} = \frac{1}{2} C(d)V^2 \tag{6.38}$$

Let us note that for d small in front of the radius of curvature of the tip, the electrostatic strength has a variation in $1/d$.

iv) Magnetic Strength

The magnetic strength acts over a long distance. With a magnetic tip, it is possible to probe properties of a surface. The magnetic interaction opened a whole domain of the microscopy in near field. Indeed, the magnetic force microscope (MFM) is a powerful research tool. It allows, among others, studying the magnetic data storage in computers or the vortices in the superconductors.

v) Capillarity Strength

Except in low temperatures and in the ultra high vacuum, the surfaces are covered with a fine layer of water which forms a water meniscus with the tip. It appears an attractive force which acts over a long distance.

B. Probe

The probe is the central element of the AFM. It is formed of a microscopic cantilever made of silicon or silicon nitride with an optic detection. The cantilever is a girder supporting a tip which is going to probe the surface. The cantilever is characterized by three parameters: the rigidity k, the frequency of resonance f_0 and the factor of quality Q. The geometry of a cantilever in the shape of gill permits it to have an important rigidity in the plane and a weak rigidity in the perpendicular axis to the probed surface. The cantilever's characteristics depend on the mode in which it will operate. The laser beam is reflected on the cantilever before reaching a detector (see Fig. 6.10). The variation of the direction of the laser beam when the cantilever interacts with the surface provides some information on the surface topography. Thus the probe is permitted to transform the effect of the strength in to an electrical signal.

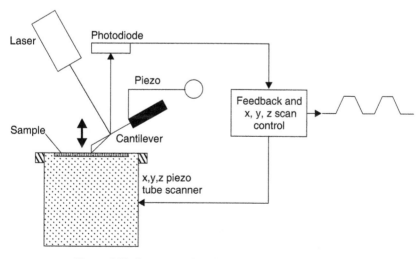

Figure 6.10. Representation of the beam-bounce deflection.

C. Operating Modes

i) Contact Mode

The tip is brought to the contact of the surface in order to make the cantilever bend. The deflection maintained is constant during the sweep and is directly given by the ratio F/k where F is the strength due to the potential of the surface and k the constant of stiffness of the cantilever. This mode of imagery is therefore at constant strength. The sign of deflection depends on the part of the surface potential which works the probe: the deflection is negative in the repulsive part and positive in the attractive part. The constant stiffness of the cantilever k must be lower than the constant of the strengths which connect the atoms of the tip and the surface. The physical interpretation of the AFM pictures in contact mode is simple: the picture is a surface $z(x,y,F = C^{te})$. The atomic resolution in contact mode was established by Giessibl and Binnig in 1992 on a surface of KBr[78].

ii) Non-contact Mode

In this mode, the tip is maintained at a distance of between 50 and 100Å. One deliberately excites the cantilever with amplitudes included between a few tenths and a few tens of nanometers. Far from the surface, the probe can be assimilated to a harmonic oscillator.

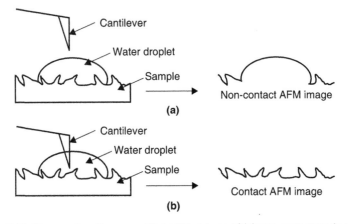

Figure 6.11. Comparative diagrams of the AFM pictures of (a) non-contact mode and (b) contact mode of a surface covered of a water droplet.

Fig. 6.8 shows that in the non-contact region, the slope of the Van der Walls curve is less steep compared to the one of the contact region. It indicates some weak deflections of the cantilever in answer to the changes of the distance tip-sample. Besides, an inflexible cantilever is necessary in order to avoid the collage of the surface of the sample under the effect of the attractive strength. The non-contact mode induced a very weak force between the tip and the sample, of the order of 10^{-12}N. This force is very useful for the study of soft or elastic samples. Another advantage is that the wafers of silicon are not contaminated by the contact of the tip. In the case of rigid samples, the obtained pictures by contact and non-contact modes can be very similar. However, if the surface of the sample is covered with a fine layer of water, the obtained pictures by the two modes will be completely different. The AFM operating in contact mode will come into the liquid layer for imaging the plane of the surface of underneath whereas in the non-contact mode, the AFM will give an image of the liquid plane (see Fig. 6.11).

iii) Intermittent Mode

In the intermittent mode, the cantilever oscillates at its resonance frequency with great amplitudes, of the order of 1000Å. The tip touches the sample at every oscillation (from where the term "intermittent" is being referenced). This mode causes less damage to the sample than the contact mode because it eliminates the lateral strengths (friction and drive strengths) between the tip and the sample. But the vertical strengths are more important than the strengths of capillarity (10^{-8}) in order to permit to the tip to immerse and emerge from the water layer. These strengths are raised so that they are capable to distort the surface of a soft or elastic material. The pictures of the intermittent mode represent a mixture of the topographic and elastic properties of the sample surface.

7.3.4. Scanning Electronic Microscope (SEM)

The operating of the scanning electronic microscope (SEM) is based on the following principle: a very thin beam of electrons sweeps the surface of a sample where there occurs an interaction with the sample. These last are picked up by a detector that controls the brilliance of a cathodic oscilloscope whose the sweep is synchronized with the one of the electrons beam. Fig. 6.12 gives a schematic representation of SEM, it includes two distinct parts: the column on the left and the cathodic screen on the right.

The electrons are produced by thermo-electronic effect from a filament of tungsten carried at a high temperature. It is in the shape of a hairpin in order to localize the broadcast by tip effect to have a punctual and brilliant source. These electrons are extracted from the Wehnelt whose the function is to focus the electrons in one point (punctual luminous source) (see Fig. 6.13). The electrons are accelerated by the difference of potential between the Wehnelt and the anode (diaphragm). The potential of acceleration V_0 (typically of 10 to 50kV) consolidates the electrons energy which forms a mono-kinetic beam.

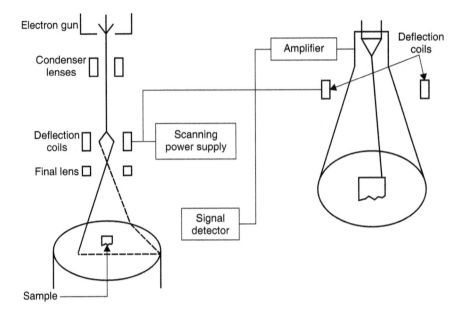

Figure 6.12. Scanning electronic microscope (SEM). It includes two distinct parts, the column on the left and the cathodic screen on the right.

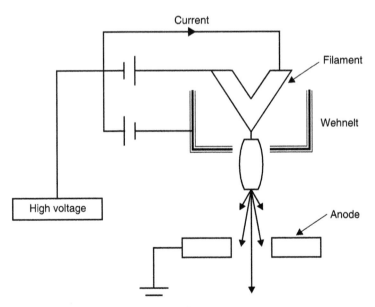

Figure 6.13. Electrons gun.

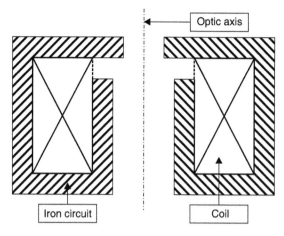

Figrue 6.14. Representation of a magnetic coil.

It is known that the trajectory of an electron in movement is deviated when it crosses an electrical or magnetic field ($F = - e\ E$) ($F = m \times dv/dt = - e\ v \times B$). In this SEM, one uses the magnetic coils which introduce very little aberrations. These coils are composed of a coil shut in an armature made of steel in order to canalize the flux. It includes an opening (see Fig. 6.14). The shape of the magnetic field lines, as an electron, will be of as much more deviated that its trajectory diverges in relation to the axis of the system, which behaves like a convergent lens. The focal length varies with the speed (energy) of electrons: a diaphragm placed close to the point of focusing permits the elimination of the electrons whose energy does not have the nominal value or that diverge too much in relation to the axis of the system. The first lens is called "condenser". It permits to control the electrons beam density by playing on its divergence. The control of the sample surface sweep is assured by deflection coils, supplied by an electrical source delivering an irregular current (deviation in x) and a current in staircase (deviation in y): the beam therefore moves on the sample surface following the lines xx' (see Fig. 6.15). A cathodic oscilloscope is synchronized with this sweep: one merely applies a current coming from the same source to the coils of deflection of the column and of the oscilloscope. There is therefore a relation between one point of the sample and one point of the screen. The obtained magnification is the ratio of the swept surfaces on the sample and the screen; it can reach 400.000. A final coil permits to control the conic shape of the beam of electrons in order to focus it precisely on the surface of the sample. Indeed, the maximal resolution (capacity to separate two points) is directly bound to the diameter of the beam at the surface. The small supplementary coils permit to correct the aberrations, in particular the astigmatism.

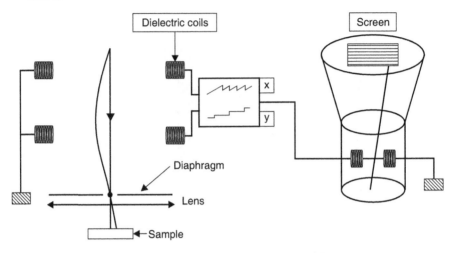

Figure 6.15. Sample surface sweep principle.

The column of the microscope is maintained under vacuum for two reasons, the first is to avoid the oxidization of the source and secondly to avoid the braking and the deviation of electrons by collision with the molecules of air. One uses for it a pump with diffusion of oil coupled with a primary pump with paddles permitting to reach $10^{-5} - 10^{-6}$Torr. The sample is introduced through an airlock to avoid breaking the vacuum in the column. Finally; the sample is placed on a micrometric turntable permitting displacements in x, y and z.

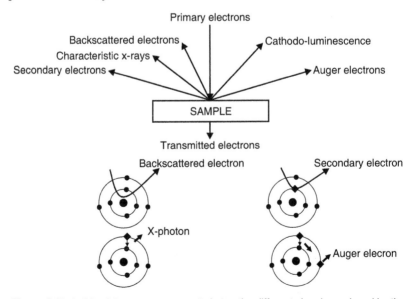

Figure 6.16. In this picture, one represents in top the different signals produced by the interaction of primary electrons with the sample and below their origins.

Fig. 6.16 summarizes the main interactions of electron-matter. They lead to the accumulation of loads at the surface; these loads are evacuated toward the earth in the case of a conductor; in the case of an insulator, their accumulation distorts the electrons beam and modifies its efficient energy, it is therefore necessary to deposit a metallic layer which is thin at the surface (gold, gold-palladium, graphite).

7.4. Elaboration Methods

We prepared $YBa_2Cu_3O_{7-\delta}$ (YBCO in short) and $Y_3Fe_5O_{12}$ (called hereafter YIG) nanoparticles using a melting process of mixtures of copper oxide (Alpha Aesar 99.7%), yttrium oxide (Rhône–Progil. SA 99.9999%) and barium peroxide (Aldrich Chemical Company. Inc 95%) for YBCO, and iron oxide (Alpha Aesar 99.9454%) and yttrium oxide (Alpha Aesar 99.99%) for YIG. In order to avoid any contamination, drastic precautions were observed. We first crushed YBCO with a new and non-contaminated ball mill. Then, we carefully washed the ball mill with an aqueous solution of nitric acid. Lastly, we crushed YIG. The mixing of YBCO and YIG was performed using several containers filled with acetone and not in the ball mill. To obtain the YBCO powder, a mixture of CuO, Y_2O_3 and BaO_2 in the stoichiometric Y-123 composition was heated at 1100°C for 24 hours and oxygenated at 500°C for another 12 hours. YIG powder was obtained by heating the mixture of Y_2O_3 and Fe_2O_3 in open air above 1350°C for 14 hours. For magnetic measurements, we used several sample holders made of plastic. Each sample holder measured 0.5 cm in diameter and 2 cm in length. Each sample had its own sample holder. Then, the sample holder was screwed onto a non-magnetic rod before introducing it in the magnetometer. This procedure has the merit of not introducing any magnetic impurities in pure YBCO thereby ruling out any risk of contamination of our samples. We produced pure YBCO and YBCO/YIG samples by pressing pure YBCO first followed by YBCO nanoparticles with different YIG concentrations (0.5, 1, 3, 9 and 20 % by weight) into disks of 5 mm of diameter and 0.5 mm of thickness. The obtained pellets were sintered at 850°C in the air for 24 hours, then at 500°C for another 12 hours and finally cooled down to room temperature in the presence of oxygen. Powder x-ray diffraction (XRD) was used to identify the different phases and purity of our samples. Their crystalline structure and phase composition were analyzed with a Siemens x-ray model D5000 diffractometer using CoK_{α} (λ = 0.1784970nm) radiation. The spectra were recorded in the region of 2θ = 10-89° with a counting time of 12s and a step of 0.04°. The unit-cell parameters were calculated and refined using the FullProf crystallographic software. The average crystallite sizes of synthesized powders were determined through XRD reflection using the well-known Scherrer equation: $D_{hkl} = 0.9\ \lambda/\beta_{1/2}\ \cos\theta$, where, D_{hkl} is the crystallite size in nm, λ is the wavelength of CoK_α radiation, $\beta_{1/2}$ is the calibrated half-width of the strongest reflection, and θ is the diffraction angle of the strongest peak. We obtained an average grain size of 80nm.

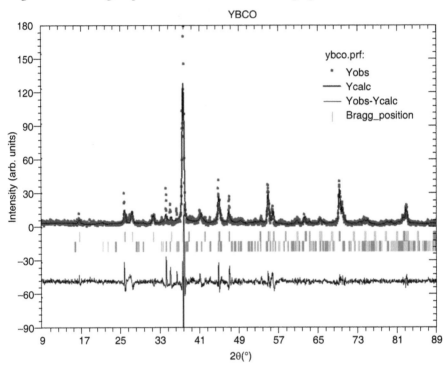

Figure 6.17. XRD pattern of YBCO sample and its FullProf refinements (dots) experimental curve (upper line) theoretical curve (lower line) difference between the above-mentioned two curves (vertical dashes) Bragg positions (after M. Gasmi, S. Khene, G. Fillion, 2013[66]).

Fig. 6.17 shows the x-ray diffraction pattern of $YBa_2Cu_3O_{7-\delta}$ that confirms the presence of this phase, of space group Pmmm, of orthorhombic structure and of unit-cell parameters $a = 0.382270$nm, $b = 0.387508$nm and $c = 1.1495180$ nm. These values are consistent with those reported in the literature[79]. Elsewhere, the diagram of Figure 6.17 also reveals the presence of the green phase Y_2BaCuO_5 (also known as 2115) in very small quantities (represented by the six small peaks at 2θ = 33.80°, 39.42°, 42.46°, 49.43°, 52.25° and 73.50°). It is worth mentioning that this green-phase generates pinning centers of vortices in the Y–123 matrix. Figure 6.18 shows the x-ray diffraction pattern of YIG. The presence of $Y_3Fe_5O_{12}$ phase of space group $Ia\bar{3}d - (O_h^{10})$ N°230 with a body-centered cubic structure and a unit-cell parameter a = 1.2349956nm is confirmed. This value is close to that reported in the literature[80]. The diagram of the fig. 6.18 also reveals the presence of impurities of $YFeO_3$ and Fe_2O_3 in small quantities (represented by the four small peaks at 2θ = 27.33°, 32.92°, 34.64° and 47.29°). For the two systems, the positions of atoms in the unit-cell were also calculated. Magnetization measurements vs. temperature T and field H were performed using a conventional magnetometer within the temperature range 4.2 – 110K and in magnetic fields up to 10T. The field was applied parallel to the pellet axis.

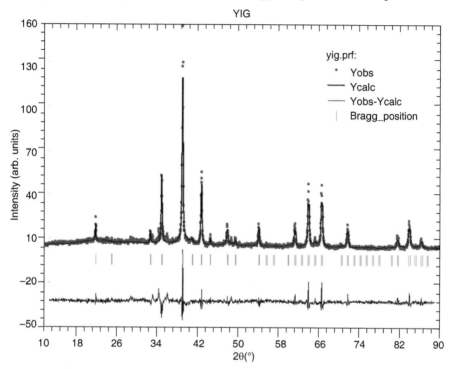

Figure 6.18. XRD pattern of YIG sample and its FullProf refinements (dots) experimental curve (upper line) theoretical curve (lower line) difference between the above-mentioned two curves (vertical dashes) Bragg positions (after M. Gasmi, S. Khene, G. Fillion, 2013[66]).

7.5. Results and Discussion

For zero-field-cooled (ZFC) magnetization measurements, samples were cooled from 110K to 4.2K in zero fields. Then, the magnetization was measured after the application of the field, while warming. The superconducting transition temperature T_C is defined as the onset temperature at which a diamagnetic signal is observed. Resulting $M(T)$ curves (see Fig. 6.19). For pure YBCO and YBCO + YIG 9 Wt. % pellets reveal T_C values of 92.37K and 89.43K, respectively. Note that $M(T)$ curves exhibit large transition widths ($\Delta T \sim 50$K) and a high value of T_C (i.e., 92.37K) for pure YBCO pellet that is a strong indicator of its purity.

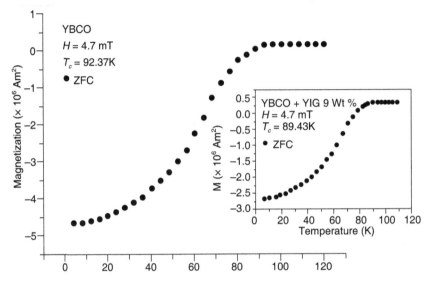

Figure 6.19. *M(T)* curves under Zero Field Cooled (ZFC) process of YBCO and YBCO + YIG 9 Wt % pellets (in the inset) (after M. Gasmi, S. Khene, G. Fillion, 2013[66]).

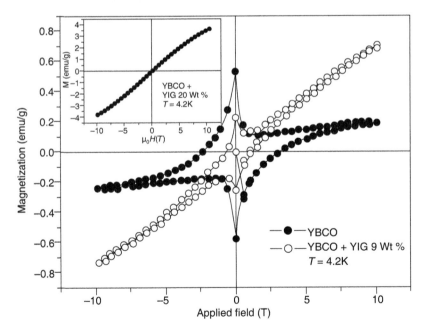

Figure 6.20. Hysteresis loops of the magnetization *M* vs. applied field *H* at 4.2K for H cycled between ±10T of YBCO and YBCO + YIG 9 Wt. % pellets. *M(H)* curve of YBCO + YIG 20 Wt. % pellet at 4.2K is shown in the inset (after M. Gasmi, S. Khene, G. Fillion, 2013[6]).

$M(H)$ curves (see Fig. 6.20) of pure YBCO and YBCO + YIG 9 Wt. % pellets at 4.2K reveal a marked hump that is centered at $H = 0$ followed by a magnetization bent shape at high fields that makes a certain angle with the field axis. From these $M(H)$ curves, it can be inferred that the low-H peak is related to some interactions that vary more rapidly with H than those controlling the high field magnetization. It is particularly interesting to notice that the $M(H)$ curve of YBCO + YIG 20 Wt. % pellet at 4.2K (see the inset of Fig. 6.20) is that of a ferromagnetic material with a coercivity H_C of 80.79×10^{-3}T and a remanence M_R of 0,0582A.m^2/kg. The weak values of H_C and M_R are due to the crystallite size reduction to the nanometer scale[81]. Fig. 6.21 shows $|M| \times |H|$ versus H where the graph reveals a hump at low field H that is inferior to a threshold field H^*. This hump is followed by a quasi-linear increase of $|M| \times |H|$ for $H > H^*$ that is identical to the one of the YIG, the well-known ferromagnetic material (see the inset of Fig. 6.21). Furthermore, this second part of the curve is nicely fitted with the Weiss and Forrer law that describes hysteresis cycles of ferromagnetic materials at high fields[82]:

$$M = M_S (1 - a/H - b/H^2 - c/H^3 - ...) + \chi_0 H \qquad (6.39)$$

where M_S is the saturation magnetization and $\chi_0 H$ denotes the spontaneous magnetization that is an increase function of temperature. The factor "a" is called "magnetic hardness". In 1948, Néel[83] showed that it resulted from non-magnetic cavities drowned in the ferromagnetic phase. He based his argument on the Lorin's experimental results on the porous ion samples obtained from powders compacted in cylinders and sintered at 850°C in H_2. Lorin's results revealed an increase of the magnetic hardness with cavities. In our pellets, we observed several cavities by SEM surface image. Néel established an approach law as a function of cavities. This law is quite complex: for fields greater than 10^5Oe, the magnetization varies as $1/H^2$ whereas for lower fields, it varies as $1/H$. The best fit of our experimental data for $H > H^*$ with the relationship 4.40 is depicted in Fig. 6.21:

$$M H = - a M_S + M_S H + \chi_0 H^2 \qquad (6.40)$$

we obtain a = 3.55T, M_S = 0.25A.m^2/kg (= 2.10 μ_B/mol) and χ_0 = 0.00316A. m^2/kg/T. Note that the $\chi_0 H$ variation versus H for 4.2, 10 and 77K clearly indicates an increase of $\chi_0 H$ with temperature which is in total agreement with the Weiss and Forrer law. For YIG, the best fit with the equation 4.40 gives a value of M_S equal to 4.75 μ_B/mol. This value is close to that reported in the literature.[84] Another interesting result consists of the graphs of Fig. 6.22 that represent the variations of $|M| \times |H|$ versus H for three YIG concentrations. We can clearly see that H^* decreases with the increase of YIG concentrations. This indicates that the ferromagnetism of YIG nanoparticles increases the number of ferromagnetic nano-domains of pure YBCO in comparison with the superconducting domains number. For a concentration of YIG of 20% in weights (see the inset of Fig. 6.22), the ferromagnetic nano-domains become preponderant in the sample; they conceal the superconducting part of the curve.

We also found that H^* decreases with temperature for both pure YBCO and YBCO/YIG samples. This phenomenon could be explained by the fact that the temperature destroys the Coopers pairs and enhances the parallel alignment of spins in the field direction. These observations are strong indicators that H^* is a transition point between the ferromagnetic phase and the superconducting phase. Moreover, the presence of the ferromagnetic phase in the superconducting material could also be inferred from the widening of the transition widths of pure YBCO and YBCO + YIG 9 Wt. % pellets in the $M(T)$ curves of Fig. 6.19. It is also clear that the particular form of hysteresis loops of Fig. 6.20 is another hint of the coexistence of superconductivity and ferromagnetism in our samples. The field dependence of $|M|\times|H|$ in both increasing and decreasing fields is shown in Fig. 6.23.

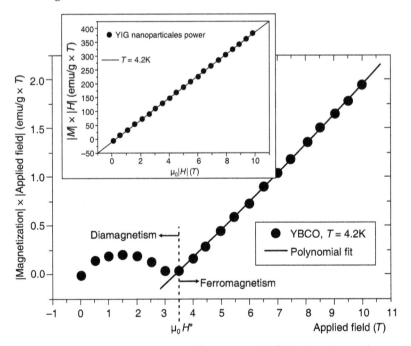

Figure 6.21. $|M|\times|H|$ vs. H of YBCO pellet at 4.2K. Field dependence of $|M|\times|H|$ of YIG nanoparticles powder at 4.2K is shown in the inset (after M. Gasmi, S. Khene, G. Fillion, 2013[66]).

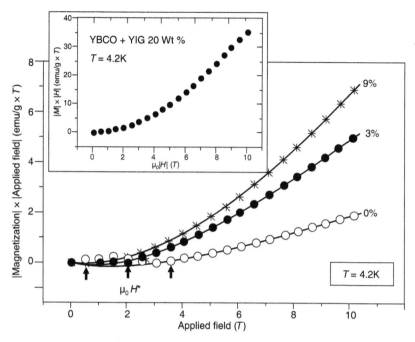

Figure 6.22. Field dependence of $|M| \times |H|$ of YBCO (0% of added YIG) and YBCO/YIG pellets for two concentrations of YIG powder (3 and 9 %) at 4.2K. $|M| \times |H|$ vs. H of YBCO + YIG 20 Wt. % pellet at 4.2K is shown in the inset (after M. Gasmi, S. Khene, G. Fillion, 2013[66]).

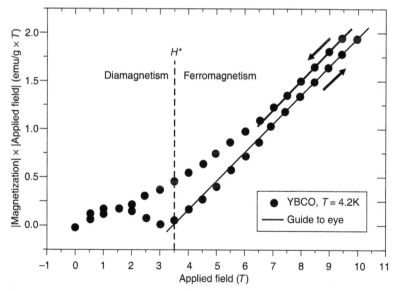

Figure 6.23. Field dependence of $|M| \times |H|$ in increasing (arrow up) and decreasing (arrow down) fields of YBCO pellet at 4.2K. Solid line is a guide to eye (after M. Gasmi, S. Khene, G. Fillion, 2013[66]).

In decreasing fields, we observe a gap in the linearity at $H = 6.5T$. For $H > H^*$, the best fit with the Weiss and Forrer law gives M_S = 0.125A.m²/kg and a value of M_S = 0.25A.m²/kg for $7T < H < 9.5T$ (linear part only) that is the same for increasing fields. However, we observed that the transition between ferromagnetism and superconducting phases in decreasing fields takes place gradually whereas it is quite abrupt for increasing fields. This means that it is easier to break the Cooper pair than to rejoin it. Ferromagnetism and superconductivity coexist in this sample. Indeed, for $H < H^*$, the spins condense to form Cooper pairs in some nano-domains of the sample whereas the rest of the nano-domains remains ferromagnetic. When the field increases, the spins in the superconducting nano-domains line up along the applied field direction to create a ferromagnetism phase that reinforces the ferromagnetic nano-domains. In decreasing fields, the spins of the superconducting nano-domains condense again to re-form Cooper pairs. One interpretation of these interesting results is to consider that the ferromagnetic nano-domains number in the sample is constant and it is the superconducting nano-domains that swing between the ferromagnetic and superconducting states under the effect of the applied field. As said previously, ferromagnetism and superconductivity have been considered antagonistic until the discovery of their coexistence in UGe_2 and $URhGe$[58,84-86]. It is well known that for conventional superconductors, magnetic fields destroy the superconductivity but in our case, the ferromagnetism and superconductivity coexistence indicates that the superconductivity is not conventional. Various theoretical models attempted to explain this coexistence by assuming a non-conventional Cooper pairs build-up mechanism. They consider that the magnetic excitations lead to the formation of Cooper pairs. These models differ from one another in the nature of these excitations and the spins orientation in the pair. Fay and Appel[87] suggest a theory based on an itinerant ferromagnetism and Cooper pairs with parallel spins (spin triplet). The mechanism of the Cooper pair formation in YBCO is still unclear. It is also possible that the phonons and the magnetic fluctuations act on this formation[88]. Using the muon spin relaxation (μSR) method on $La_{2-x}Sr_xCuO_4$ single crystals with Sr content x = 0.15, 0.166, 0.176, 0.196, 0.216, 0.24, and 0.33, Sonier et al.[89] presented results that lend support to electronic band calculations that predict the occurrence of weak localized ferromagnetism at high doping. Barbiellini and Jarlborg[90] suggested that the superconductivity and ferromagnetism in singlet superconductors such as cuprates do not exist uniformly but are separated in nano-domains. Their works confirms that the local spin-density approximation electronic structure of $La_{2-x}Ba_xCuO_4$ in the overdoped regime is consistent with weak ferromagnetism appearing locally around clusters of high Ba concentration. These latter works validate in fact the hypothesis of Kopp et al.[91] that ferromagnetism order and superconductivity are competing in cuprates such as $Tl_2Ba_2CuO_{6+d}$ and $La_{2-x}Sr_xCuO_4$ that can be overdoped to the edge of the superconducting dome. As for the singlet-type superconductor such as cuprates, the uniform coexistence of the two orders is difficult to understand, the idea of the superconductivity and ferromagnetism coexistence in separate nano-domains may well explain our results. Indeed, as shown in the fig. 4.22, the threshold field H^* decreases when the number of YIG nanoparticles increases. This indicates that the ferromagnetism of YIG nanoparticles increases the number of ferromagnetic nano-domains in pure YBCO in comparison with the superconducting nano domains number. For a concentration of YIG of 20% (see the inset of Fig. 6.22), the ferromagnetic nano domains become preponderant in the sample and as a consequence, the superconducting part of the curve becomes imperceptible. For a fixed ferromagnetic nano-domains number (pure YBCO pellet for instance), a higher magnetic field beyond H^* destroys the superconductivity in superconducting nano-domains and leads to a phase transition to a ferromagnetic state. The number of the ferromagnetic nano-domains is stationary in the sample; it does not vary as a function of the field. These nano-domains are on the surface of the sample. When the value of the applied magnetic field increases beyond H^*, the superconducting nano-domains that are inside the sample go from the superconducting state to the ferromagnetic state and vice versa when the field decreases. This interpretation finds its justification in the results of Hasanain et al.[64] Zhu et al.[65] who mentioned that ferromagnetism in nano particles of YBCO occurs on the surface whereas the superconductivity occupies the volume of the sample. The deepened study of the competition between the superconductivity and ferromagnetism in high T_c superconductors contributes to explain the mechanism of the electrons pairing in these materials. In conclusion, coexistence of ferromagnetism and superconductivity in nanosized powders of YBCO has been established at 4.2K. For H less than a threshold field H^* which is a decreasing function of temperature and added YIG percentage, there is a predominance of the superconducting nano-domains in the sample. When the field increases beyond H^*, the spins into the superconducting nano-domains line up along the applied field direction to create a ferromagnetism phase that reinforces the ferromagnetic nano domains that are on the sample surface. In decreasing fields, the spins of superconducting nano-domains that occupy the sample volume condense again to form Cooper pairs. Superconductivity and ferromagnetism in YBCO do not exist uniformly but are phase-separated in nano-domains.

8. Vortex Pinning by the Spins

8.1. Introduction

In the previous paragraph (see paragraph 7 of this chapter), we demonstrated the coexistence of ferromagnetism and superconductivity in nanosized powders of YBCO at 4.2K. We established that for H less than a threshold field H^* which is a decreasing function of temperature and added YIG percentage, there is a predominance of superconducting nano domains in the sample. When the field increases beyond H^*, spins into the superconducting nano-domains line up along the applied field direction to create a ferromagnetic phase that reinforces the ferromagnetic nano domains that are on the sample surface. In decreasing fields, the spins of superconducting nano-domains that occupy the sample volume condense again to form Cooper pairs. Superconductivity and ferromagnetism in YBCO do not exist uniformly but are phase-separated in nano-domains. In the present work, we are interested in the mechanisms of vortices trapping in this system. Indeed, the properties of superconductor-ferromagnet systems have been the object of many theoretical and experimental investigations during the last few years[92]. In addition to electronic interactions[93-98], it has been found that the ferromagnetic order can directly influence the superconducting phase properties[99-100]. The commercial use of a superconductor is limited by its ability to pin vortices. Several researchers have looked into the pinning potential through the incorporation of ferromagnetic particles within a superconducting matrix[101-105]. Although some pinning enhancements have been observed, the pinning interaction nature is not well understood, partly because of the difficulty of achieving a sufficiently fine dispersion of the appropriate particles[106-107].

8.2. Elaboration Methods

The elaboration methods are those of the previous study (see paragraph 7 of this chapter).

8.3. Results and Discussion

$M(H)$ curves of pure YBCO at 4.2, 10, 60 and 77K are shown in Fig. 6.24. These curves reveal a marked hump that is centered at $H = 0$. This peak is followed by a magnetization bent shape at high fields that makes a certain angle with the field axis. From these $M(H)$ curves, it can be inferred that the low-H peak is related to some interactions that vary more rapidly with H than those controlling the high field magnetization. Fig. 6.25 shows $|M| \times |H|$ versus H where the graph reveals a hump at low field H that is inferior to a threshold field H^*. This hump is followed by an increase of

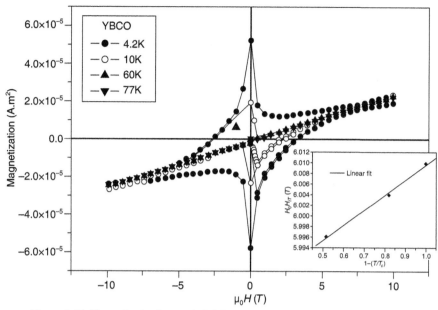

Figure 6.24. Magnetization loops of YBCO at temperatures of 4.2, 10, 60, and 77K. This figure shows that for all temperatures, the H-peak is located at $H = 0K$. In the inset, we show a linear variation of $\mu_0 H_{irr}$ as a function of $1 - (T/T_c)^4$. H_{irr} is the field at which the increasing and decreasing branches of hysteric cycle become distinct (after S. Khene, M. Gasmi, G. Fillion, 2015[107]).

$|M| \times |H|$ for $H > H^*$. Furthermore, this second part of the curve is nicely fitted with the Weiss and Forrer law (see Equation 6.39) that describes hysteresis cycles of ferromagnetic materials at high fields.[82–28]

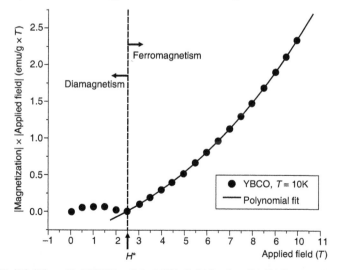

Figure 6.25. $|M| \times |H|$ vs. H of YBCO pellet at 10K. Solid line is a fit with the equation 6.40 (after S. Khene, M. Gasmi, G. Fillion, 2015[107]).

The best fit of our experimental data for $H > H^*$ with the equation 4.40 is depicted in Fig. 6.25. We obtained $a = 4.39T$, $M_s = 0.059A.m^2/kg$ and $\chi_0 = 0.020A.m^2/kg/T$. We found that the $\chi_0 H$ variation versus H for 4.2, 10, 60 and 77K clearly indicates an increase of $\chi_0 H$ with temperature which is in total agreement with the Weiss and Forrer law. Another interesting result consists of the graphs of the figure 4.26 that represent the variations of $|M| \times |H|$ versus H for four temperatures 4.2, 10, 60 and 77K. We can clearly see that H^* decreases with temperature for pure YBCO. This phenomenon could be explained by the fact that the temperature destroys the Cooper pairs and enhances the parallel alignment of spins in the field direction. These observations are strong indicators that H^* is a transition point between the ferromagnetic phase and the superconducting phase (see the phase diagram in the inset of Fig. 6.26).

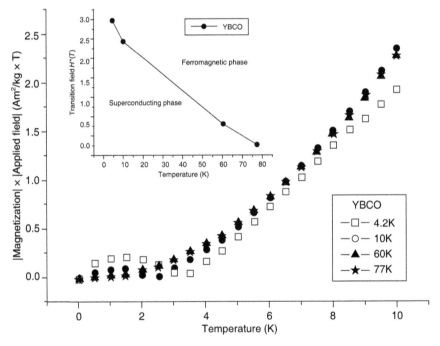

Figure 6.26. Field dependence of $|M| \times |H|$ vs. T of YBCO pellet for four temperatures. Phase diagram of YBCO pellet is shown in the inset (after S. Khene, M. Gasmi, G. Fillion, 2015[10]).

$|M| \times |H|$ variations versus H for four YIG concentrations have been also studied[66]. We have found that H^* decreases with the increase of YIG concentrations. This indicates that the ferromagnetism of YIG nanoparticles increases the number of ferromagnetic nano-domains of pure YBCO in comparison with the superconducting domains number. For a concentration of YIG of 20% in weight, the ferromagnetic nano-domains become preponderant in the sample; they conceal the superconducting part

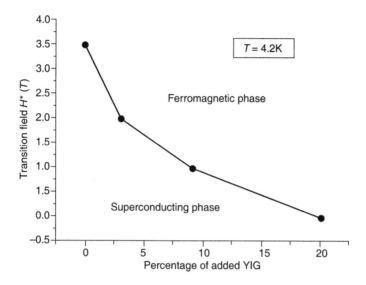

Figure 6.27. Transition field H^* of YBCO/YIG system for different proportions of added YIG at 4.2K (after S. Khene, M. Gasmi, G. Fillion, 2015[107]).

of the curve. Fig. 6.27 gives the transition field variation H^* of YBCO/YIG system as a function of added YIG percentage at 4.2K. The inter-granular apparent magnetic current values J_C have been calculated using the simplified Bean model:[108]

$$J_C = 15 \, (M^+ - M^-)/R \qquad (6.41)$$

where J_C is the critical current, M^+ and M^- are the superior and inferior legs of the cycle and R, the pellet radius. Note that this formula gives proportionality between J_C and ΔM. $J_C(H)$ curve of YBCO pellet at 10K deduced from the hysteresis loop is shown in Fig. 6.28.

It can be seen a rapid decreasing of J_C at low fields ($H < H^*$) followed by a slackening of this decreasing at high fields ($H > H^*$). We have found that the dependence:

$$J_C(H) = A_1 \, exp(-H/h_1) + A_2 \, exp(-H/h_2) \qquad (6.42)$$

fits remarkably well our experimental data (solid line in the figure with $A_1 = 1.56 \times 10^6 \text{A/m}^2$, $h_1 = 0.18\text{T}$, $A_2 = 1.04 \times 10^6 \text{A/m}^2$ and $h_2 = 1.69\text{T}$) indicating the existence of two pinning mechanisms: the first term of the right hand side corresponds to the low-H domain whereas the last one describes the high field behavior. In order to interpret the exponential decay of second order of J_C vs. H, we propose a two currents model:

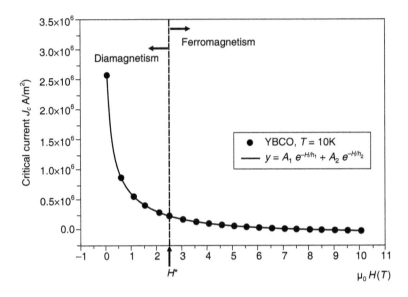

Figure 6.28. $J_c(H)$ curve of YBCO pellet at 10K. Solid line is a fit with Equation 6.42 (after S. Khene, M. Gasmi, G. Fillion, 2015[107]).

$$J_C(H) = J_{lr} (H/H_{lr}) + J_{sr} (H/H_{sr}) \qquad (6.43)$$

where the subscripts "lr" and "sr" stand for long range and short range interactions of vortex lines, respectively. In other terms, the low-H peak of the hysteresis cycle would correspond to some long interactions which drop very rapidly with H than that controlling the high field magnetization. The main feature of Fig. 6.24, in the low field region, is the appearance of a peak centered at $H = 0$ in both field increasing and decreasing branches of hysteresis curves for four temperatures, 4.2, 10, 60 and 77K. The low-H peak is unexpected and its origin remains a subject of research. It has received many explanations[109-112]. In our case, the fact that it is always centered at $H = 0$ indicates that it could be induced by some kind of surface pinning. Indeed, the surface works as a barrier for flux penetration in and out of the sample. Such a surface barrier appears due to the interaction between the vortex and its "mirror image" near the surface[113] and is rather strong in HTSC due to small values of ξ. However, even small surface imperfections destroy the mirror image effect and reduce the barrier[114]. The sharp kink at $H \sim 0.5T$ in the field increasing branch of the virgin curve characterizes the first flux penetration field H_p and its sharpness indicates the weak contribution of bulk pinning and the good quality of the sample (see Fig. 6.29).

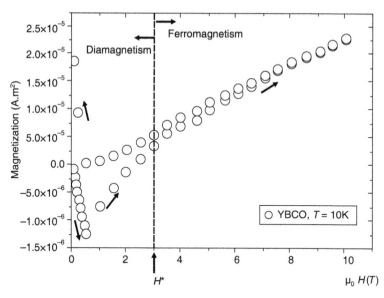

Figure 6.29. Initial branch of the hysteresis cycle of YBCO at $T = 10K$ (after S. Khene, M. Gasmi, G. Fillion, 2015[107]).

During the process of field decrease and in the low field region ($H < 2.5T$, superconducting region), the magnetization M is approximately zero below the peak and its variation is relatively flat before increasing again at $H = 0$. This $M \sim 0$ value is the surface barriers fingerprint[115–118]. On the decreasing field, there is always an effective barrier preventing the vortex to exit the superconductor, until M reaches a positive value[119]. The second current term of the right hand side of Equation 6.42 corresponds to a spins-vortex interaction. Indeed, in the high field region (above H^*) where the ferromagnetic phase predominates over the diamagnetic one, the vortex motion is slowed down by a great distribution of well-ordered spins. During its displacement, the vortex misleads the spins and causes an increase of magnetization[120]. It is worth noting at the outset that the interactions vortex-defect and spin-defect are negligible in the presence of magnetic interactions[121-126]. In our case, direct coupling between vortices and individual ferromagnetic nano-domains is precluded by the ferromagnetic nano-domain size and density which is significantly smaller than the penetration depth λ. Indeed, the superconducting nano-domain density and size are such for YBCO the volume λ^3 will contain hundreds of ferromagnetic nano-domains (see Fig. 4.30). The vortex displacement will change the effective field on the ferromagnetic nano-domains at both the new and the old site. Any hysteresis in the magnetization of YBCO will mean that this displacement requires energy input. This new form of flux pinning is associated with magnetic hysteresis. Indeed, conventional core pinning of vortices is due to the savings in condensation energy associated with the vortex core which arise when the core is situated within a non-superconducting inclusion. The energy per unit length of vortex associated with core pinning is different of magnetic pinning which is associated with the Zeeman energy per unit length of the particles. In our demonstration, we consider a vortex represented by a cylindrical flux of radius λ, and so the magnetic (Zeeman) energy per unit length of the vortex is given by[106]:

$$\varepsilon = \pi\lambda^2 B_V M \tag{6.44}$$

where B_V is the flux density within the vortex and M the mean magnetization within the vortex.

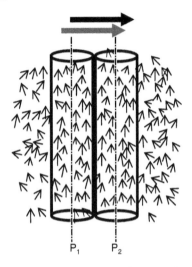

Figure 6.30. Schematic diagram of the interaction between the moments of an array of ferromagnetic nano-domains and the field within a vortex in movement from site P_1 to site P_2 (after S. Khene, M. Gasmi, G. Fillion, 2015[107]).

We started by assuming that a field sufficient to exceed $\mu_0 H_{C2}$ was applied and then progressively reduced to zero. In any field, a vortex moving from one position P_1 to another P_2 reduces the magnetization at the original core location from M_1 to M_2 and increases the magnetization at the new location from M_2 to M_3. Therefore the change in energy per unit length for a vortex moving from site P_1 to site P_2 is given by:

$$\Delta\varepsilon = \pi\lambda^2 B_V (M_3 - M_1) \tag{6.45}$$

An isolated vortex in a current $\mathbf{J}$ is subjected to a force per unit volume, called Lorentz force $\mathbf{f} = \mathbf{J} \times \Phi_0$. The lattice is subjected to a force density per unit volume $\mathbf{F} = \mathbf{J} \times n\Phi_0 = \mathbf{J} \times \mathbf{B}$, n is the number of vortices per area. This force tends to set in motion the lattice of flux lines. If the vortices can move freely, it is not possible to pass a current above $\mu_0 H_{C1}$ without energy dissipation. Therefore, to achieve a finite critical current, we would have to pin the vortices i.e., to find mechanisms or geometries which prevent flux line motion. Let us call F_p, the average pinning force density which prevents the lattice moving. If $F < F_p$, the lattice will not move and we have a non-dissipative current. If $F > F_p$, the lattice will move and we have the so-called flux flow regime. When $F = F_p$, we have the critical regime. J_C, the critical current of the material at zero temperature, is defined by $F_p = J_C B$ or for one vortex $f_p = J_C \Phi_0$. Equation 6.45 can be related to the pinning force per unit length by $f_p = \Delta\varepsilon/\lambda = \pi\lambda B_V (M_3 - M_1) = \Delta J_C \Phi_0$, ΔJ_C is the critical current enhancement, which is therefore given by $\Delta J_C = f_p/\Phi_0 = \pi\lambda B_V (M_3 - M_1)/\Phi_0$. It turns out that the most stable lattice of vortices is the triangular lattice. In this configuration, the distance a_0 between two vortices is given by $\mu_0 H_{irr} = (2\,\Phi_0)/(\sqrt{3}\,a_0^2)$. H_{irr} is the field at which the increasing and decreasing branches of hysteric cycle become distinct. Since the magnetic hysteresis pinning is entirely magnetic in origin, the magnitude of f_p for this pinning will depend on the inhomogeneity of the field in the mixed state. Therefore, it is only expected to be significant for flux densities $\mu_0 H_{irt}$ given by $\Phi_0\mu_0 H_{irr} \geq (\sqrt{3}\,a_0^2)/2$. High-$T_C$ superconductors are of a type II character. Below $\mu_0 H_{C1}$, the external magnetic field is excluded from the bulk of the material by a persistent supercurrent in the surface region. The supercurrent induces a magnetic field which exactly cancels the applied field existing inside the material. The depth of this supercurrent carrying layer is the penetration depth λ. The external field penetrates into the superconductor in an exponentially decreasing manner. For BCS superconductors; the penetration depth λ has a temperature dependence given by the relation $\lambda(T) = \lambda_0 [1 - (T/T_c)^4]^{-1/2}$. This relation is in good agreement with experimental results in conventional superconductors. Experimental results in high-T_C materials[127] show a behavior in agreement with the above relation. Taking $a_0 \sim 2\lambda$, we finally obtain:

$$\mu_0 H_{irr} = [\Phi_0/(2\sqrt{3}\,\lambda_0^2)]\,[1 - (T/T_c)^4] \tag{6.46}$$

Eq. 6.46 implies that the plot $\mu_0 H_{irr}$ vs. $1 - (T/T_c)^4$ must be linear and should have a slope of $\Phi_0/(2\sqrt{3}\,a_0^2)$. This is shown in the inset of the figure 4.24 from which $\lambda_0 = 1445 \pm 20$Å. This value of λ_0 is in agreement with the literature values.[128–138]

REFERENCES

1   J.C. Maxwell, A Treatise on Electricity and Magnetism, 3 rd, Ed. (Oxford), p. 258, 1892.
2   J. Silcox and R.W. Rollins, Appl. Phys. Lett. **2**, 231 (1963); Rev. Phys. **36**, 52 (1964).
3   J. Friedel, P.G. De Gennes and J. Matricon, Appl. Phys. Lett. **2**, 119 (1963).
4   R. Labusch, Physics Lett. **22**, 9 (1966); Phys. Stat. Sol. **19**, 715 (1967); Phys. Rev. **170**, 470 (1968); Phys. Stat. Sol. **32**, 439 (1969).
5   A.M. Campbell and J.E. Evetts, Adv. Phys. **21**, 129 (1972).
6   E.H. Brandt, J. Low Temp. **26**, 709 (1976); J. Low Temp. Phys. **26**, 263 (1977); Phys. Rev. Lett. **56**, 1381 (1986).
7   E.H. Brandt, Concise Enyclopedia of Magnetic and Superconducting Materials, ed. J. Evetts, Pergamon Press, p. 159, 1992.
8   G. Blatter, M.V. Feigel'man, V.B. Geshkenbein, A.I. Larkin and V.M. Vinokur, Vortices In Hight Temperature Superconductors, Rev. Mod. Phys. **66**, 1125 (1994).
9   (a) A.M. Schönenberger, V.B. Geshkenbein and G. Blatter (cited by (b) A.M. Ettouhami, Thesis, University Joseph Fourier-Grenoble 1, 1994.
10 C.R. Hu, Phys. Rev. **B 6**, 1756 (1972).
11 J. Friedel, P.G. de Gennes and J.Matricon, Appl. Phys. Lett. **2**, 119 (1963).
12 H.T. Coffey, Phys. Rev. **166**, 447 (1968).
13 W.A. Fietz and W.W. Webb, Phys. Rev. **178**, 657 (1969).
14 R. Labush, Crystal Lattice Deffects **1**, 1 (1969).
15 H. Ullmaier, Irreversible Properties of Type II Superconductors, Springer Tracts in Modern Physics **76** (1975).
16 A.I. Larkin and Yu.N. Ovchinnikov, J. Low Temp. Phys. **34**, 409 (1979).
17 H.R. Hilzinger and H. Kronmüller, J. Magn. Mat. **2**, 11 (1976).
18 P.A. Lee and T.M. Rice, Phys. Rev. **B 19**, 3970 (1979).
19 E.H. Brandt, Physica **C 195**, 1 (1992); Int. J. Mod. Phys. **B 5**, 751 (1991); E.H. Brandt and U. Essmann, Phys. Status Solidi **B 144**, 13 (1987).
20 M. Ousséna, S. Senoussi and G. Collin, Europhys. Lett. **4**, 625 (1987).
21 C.N. Chu, P.H. Hor, R.L. Meng, L. Gao, Z.L. Huang and T.Q. Wang, Phys. Rev. Lett. **58**, 405 (1987).
22 P. Chaudhary, R.H. Koch, R.B. Laibowitz, T.R. Mc Guire and R.J. Gambino, Phys. Rev. Lett. **58**, 2684 (1987).
23 S. Senoussi, M. Ousséna, G. Collin and I. Campbell, Phys. Rev. **B 37**, 9792 (1988).
24 O. Laborde, J.L. Tholence, P. Lejay, A. Sulpice, R. Tournier, J.J. Capponi, C. Michel and J. Provost, Solid State Commun. **63**, 877 (1988).
25 P.H. Kes, J. Van Den Berg, C.J. Van Der Beek, J.A. Mydosh, L.A. Roaeland, A.A. Menovesky, K. Kadowaki and F.R. De Boer, Proc. of the 1st Latino-American Conference on High Temp. Superconductivity, Brazil, 4-6 May (1968); World Scientific Co, Singapore, p. 239.
26 M.M. Fang, D.K. Finnemore, D.E. Farrell and N.R. Bansal, Cryogenics **29** n° 34 (supplément), 347 (1989).
27 M. Naito, A. Matsuda, K. Kitazawa, S. Kambe, I. Tanaka and H. Kojima, Phys. Rev. **B 41**, 4823 (1990).
28 C. Aguillon, D.G. Mc Cartney, P.Regnier, S. Senoussi and G.J. Tatlockt, J. Appl. Phys. **69**, 8261 (1991).
29 V.V. Moschalkov, A.A. Zhukov, D.K. Petrov, V.I. Voronkova and V.K. Yanovsku, Physica **C 166**, 185 (1990).
30 V.V. Moschalkov, A.A. Zhukov, V.D. Kuznetsov, V.V. Mellushko and L.I. Leonyuk, Pisma, Zh. Eksp. Teor. Fiz. **50**, 81 (1989).
31 H. Raffi, S. Labdi, O. Laborde and P. Monceau, Physica **B 165–166**, 1423 (1990).
32 C. Quitman, U. Ebels, P.C. Splittgerber, Hünnekes and G. Güntherodt, Physica **B 165–166**, 1143 (1990).
33 J.N. Eckstein and I. Bozovic, Appl. Lett. **57**, 1049 (1990).

34 P. Gierllowski, J. Gorecka, R. Sobolewski and S.J. Lewandowski, Physica B **165–166**, 1485 (1990).
35 M.F. Tai, Y.K. Tu and I.C. Jou, Physica B **165–166**, 1401 (1990).
36 D.E. Morris, M.R. Chandrachood and A.P.B. Sinha, Physica C **175**, 156 (1991).
37 P. Manuel, C. Aguillon and S. Senoussi, Physica C **177**, 281 (1991).
38 D. Neerinck, K. Temst, M. Baert, E. Osquiguil, C. Van Haesendonck, Y. Bruynseraede, A. Gilabert and Ivan K. Shuller, Phys. Rev. Lett. **67**, 2577 (1991).
39 H.C. Kanithi, L.R. Motowildo, G.M. Ozeryansky, D.W. Hazlton and B.A. Zeitlin, IEEE Trans. Magn. **25**, 2204 (1989).
40 S. Senoussi, J. Phys. III France **2**, 1041 (1992).
41 M. Däumling, J.M. Seuntjens and D.C. Larbalestier, Nature **346**, 332 (1990).
42 B. Roas, L. Schultz and G. Saeman-Ichenko, Phys. Rev. Lett. **64**, 479 (1990).
43 B. Janossy, H. Gu, R. Cabanel and L. Fruchter, Physica C **193**, 344 (1992).
44 C.P. Bean and J.D. Livingston, Phys. Rev. Lett. **12**, 1 (1964).
45 H. Ulmaier (Springer Tracs in Modern Physics, New York) **76**, 1 (1975).
46 D. Feinberg and A.M. Ettouhami, Int. J. of Mod Phys. B **7**, 2085 (1993).
47 S. Khene, Physica B 349 (2004) 227.
48 P.A. Mansky, P.M. Chaikin and R.C. Haddon, Phys. Rev. Lett. **70**, 1323 (1993).
49 W.K. Kwok, U. Welp, V.M. Vinokur, S. Flesher, J. Downey and G.W. Crabtree, Phys. Rev. Lett. **67**, 390 (1991).
50 G. Blatter, V.B. Geshkenbein and A.I. Larkin, Phys. Rev. Lett. **68**, 875 (1992).
51 A.M. Schönenberger, V.B. Geshkenbein and B. Blatter, cited in the reference [60].
52 M.V. Feigel'man, V.B. Geshkenbein and A.I. Larkin, Physica C **167**, 177 (1990).
53 P.W. Anderson, Phys. Rev. Lett. 9, 309 (1962); P.W. Anderson and Y.B. Kim, Rev. Mod. Phys. **36**, 39 (1962).
54 M.V. Feigel'man, V.B. Geshkenbein, A.I. Larkin and V.M. Vinokur, Phys. Rev. Lett. **63**, 2303 (1989).
55 Y. Yeshurun and A.P. Malozemoff, Phys. Rev. Lett. **60**, 2202 (1988).
56 B. Janossy, Thesis, University Paris-south, Orsay, France, 1993.
57 A. Sundaresan, C.N.R. Rao, Nano Today 4, 96–106 (2009).
58 S.S. Saxena, P. Agarwal, K. Ahilan, F.M. Grosche, R.K.W. Hasselwimmer, M.J. Steiner, E. Pugh, I.R. Walker, S.R. Julian, P. Monthoux, G.G. Lonzarich, A. Huxley, I. Sheikin, D. Braithwaite and J. Flouquet, Nature **406**, 587–592 (2000).
59 I.S. Elfimov, S. Yunoki and G.A. Sawatzky, Phys. Rev. Lett. **89**; 216403–216407 (2002).
60 A. Zywietz, J. Furthmuller and F. Bechstedt, Phys. Rev. B **62**; 6854–6857 (2000).
61 A. Shipra, A. Gomathi, A. Sundaresan and C.N.R. Rao, Solid State Commun. **142**, 685–688 (2007).
62 A. Sundaresan and C.N.R. Rao, Solid State Commun. **149**, 1197–1200 (2009).
63 A. Sundaresan, R. Bhargavi, N. Rangarajan, U. Siddesh and C.N.R. Rao, Phys. Rev. B **74**, 161306–161310 (2006).
64 S.K. Hasanain, N. Akhtar and A. Mumtaz, J. Nanopart. Res. **13**, 1953–1961 (2010).
65 Z. Zhu, D. Gao, C. Dong, G. Yang, J. Zhang, J. Zhang, Z. Shi, H. Gao, H. Luo and D. Xue, Phys. Chem. Chem. Phys, **14**, 3859–3863 (2012).
66 M. Gasmi, S. Khene and G. Fillion, J Phys Chem Solids **74**, 1414–1418 (2013).
67 E.F. Bertaut and F. Forrat., C.R. Acad. Sci. Paris **242**, 382 (1956); E.F. Bertaut and F. Forrat, C.R. Acad. Sci. Paris **244**, 96 (1956).
68 S. Geller and M.A. Gilleo, Acta. Crist. 10, 239 (1957); M.A. Gilleo and S. Geller, Phys. Rev. **110**, 73 (1958).
69 H.S. Yoder and M.L. Keith, Am. Mineralogist **36**, 519 (1951).
70 Space-group symmetry, in International tables for crystallography, edited by T. Hahn, volume A, Dordrecht/Boston/London, kluwer academic publishers edition (2002).
71 Z.A. Kazei, N.P. Kolmakova, P. Novak and V.I. Sokolov, Magnetic properties of non-metallic inorganic compounds based on transition elements, in LANDOLT-BÖRNSTEIN Group III: Condensed Matter, edited by H.P.J. Wijn, volume 27e, Berlin Heidelberg,

springer-verlag edition, ISBN 3-540-53963-8, (1991).

72 Bouguerra, G. Fillion, E.K. Hlil and P. Wolfers, Journal of Alloys and Compounds **442**, 231–234 (2007).

73 Y.H. Jeon, J.W. Lee, J.H. Oh, J.C. Lee and S.C. Choi, Phys. Status Solidi A **201** (8), 1893 (2004); T.Y. Kim, Y. Yamazaki and T. Hirano, Phys. Status Solidi B **241** (7) 160 (2004).

74 C.Y. Tsay, C.Y. Lin, K.S. Liu, I.N. Lin, L.J. Hu and T.S. Yeh, J. Magn. Magn. Mater. **239**, 490 (2002).

75 Ph Lethuillier, Louis Néel Laboratory, CNRS, Grenoble, Personal communication(1996).

76 G. Fillion, Louis Néel Laboratory, CNRS, Grenoble, Personal communication (1991).

77 P. Weiss, J. Phys. **4**, 473 (1905).

78 F.J. Giessibl and G. Binnig. True atomic resolution on KBr with a low-temperature atomic force microscope in ultrahigh vacuum. Ultramicroscopy **42–44**, 281 (1992).

79 T. Hahn, (ed.): International Tables for Crystallography, Vol. A; Space-Group Symmetry. Kluwer Academic, Dordrecht/Boston/London, 2002.

80 M. Rajendran, S. Deka, P.A. Joy and A.K. Bhattacharya, J. Magn. Magn. Mater. **301**, 212–219 (2006).

81 S. Alleg, S. Azzaza, R. Bensalema, J.J. Sunol, S. Khene and G. Fillion, J. Alloys Compd. **482**, 86–89 (2009).

82 A. Herpin, Theory of magnetism, Presses universitaires de France, Saclay, 1968.

83 L. Néel, J. Phys. Radium **9**, 184–192 (1948).

84 F.W. Harrison, J.F.A. Thompson and K. Tweedale, Proceeding ICM Nottingham, 664–665 (1964).

85 D. Aoki, A. Huxley, E. Ressouche, D. Braithwaite, J. Flouquet, J.P. Brison, E. Lhotela and C. Paulsen, Nature **413**, 613–616 (2001).

86 T. Akazawa, H. Hidaka, H. Kotegawa, T.C. Kobayashi, T. Fujiwara, E. Yamamoto, Y. Haga, R. Settai and Y. Ōnuki, Physica B **359-361**, 1138–1140 (2005).

87 D. Fay, Phys. Rev. B **22**, 3173–3182 (1980).

88 T. Kontos, M. Apriti, J. Lesueur, X. Grisona and L. Dumoulin, Phys. Rev. Lett. **93**, 137001–137005 (2004).

89 J.E. Sonier, C.V. Kaiser, V. Pacradouni, S.A. Sabok-Sayr, C. Cochrane, D.E. MacLaughlin, S. Komiya and N.E. Hussey, Proc. Natl. Acad. Sci. USA **107**, 17131–17134 (2010).

90 B. Barbiellini and T. Jarlborg, Phys. Rev. Lett. **101**, 157002–157006 (2008).

91 A. Kopp, A. Ghosal and S. Chakravarty, Proc. Natl. Acad. Sci. U.S.A. **104**, 6123–6127 (2007).

92 A.I. Buzdin, Rev. Mod. Phys. **77**, 935–976 (2005); I.F. Lyuksyutov and V.L. Pokrovsky, Adv Phys **54** (1), 67–136 (2005).

93 V.V. Ryazanov, V.A. Oboznov, A.Yu. Rusanov, A.V. Veretennikov, A.A. Golubov and J. Aarts, Phys Rev Lett **86**, 2427–2430 (2001).

94 T. Kontos, M. Aprili, J. Lesueur, F. Genêt, B. Stephanidis and R. Boursier, Phys Rev Lett **89**, 137007–137011 (2002).

95 A.Y. Rusanov, S. Habraken and J. Aarts, Phys Rev B **73**, 060505-060509 (2006).

96 A.J. Drew, S.L. Lee, D. Charalambous, A. Potenza, C. Marrows, H. Luetkens, A. Suter, T. Prokscha, R. Khasanov, E. Morenzoni, D. Ucko and E.M. Forgan, Phys Rev Lett **95**, 197201–197205 (2005).

97 A.Yu. Rusanov, M. Hesselberth, J. Aarts and A.I. Buzdin, Phys. Rev. Lett. **93**, 057002–057006 (2004).

98 R.J. Kinsey, G. Burnell and M.G. Blamire, IEEE Trans Appl Supercond, vol. **11**, 904–907 (2001).

99 I.F. Lyuksyutov and V. Pokrovsky, Phys Rev Lett **81**, 2344–2347 (1998).

100 M. Lange, M.J. Van Bael, Y. Bruynseraede and V.V. Moshchalkov, Phys Rev Lett **90**, 197006–197010 (2003).

101 N.D. Rizzo, J.Q. Wang, D.E. Prober, L.R. Motowidlo and B.A. Zeitlin, Appl Phys Lett **69**, 2285–2288 (1996).

102 A. Snezhko, T. Prozorov and R. Prozorov, Phys Rev B **71**, 024527–024533 (2005).
103 M.J. Van Bael, K. Temst, V.V. Moshchalkov and Y. Bruynseraede, Phys Rev B **59**, 14674–14679 (1999).
104 C.C. Koch and G.R. Love, J Appl Phys **40**, 3582–3588 (1969).
105 T.H. Alden and J.D. Livingston, J Appl Phys **37**, 3551–3557 (1966).
106 A. Palau, H. Parvaneh, N.A. Stelmashenko, H. Wang, J.L. Macmanus-Driscoll and M.G. Blamire, Phys Rev Lett **98**, 117003–117007 (2007).
107 S. Khene, M. Gasmi and G. Fillion, Journal of Magnetism and Magnetic Materials **373**, 188–194 (2015).
108 S. Senoussi, J. Phys. III France **2**, 1041–1257 (1992).
109 C.C. Silvan and M.E. Mchenry, IEEE Trans Appl Supercond, Vol. **7**, n° 2, 1596–1599 (1997); S. Senoussi, C. Aguillon and P. Manuel, Physica C **175**, 202–214 (1991).
110 T.H. Johansen, D.V. Shantsev, M.R. Koblischka, Y.M. Galperin, P. Naleka and M. Jirsa, Physica C **341-348**, 1443–1444 (2000).
111 Z.X. Shi, H.L. Ji, X. Jin, J.R. Jin, X.X. Yao, X.S. Rong, Y.M. Li, H.T. Peng, X.R. Long and C.R. Peng, Physica C **231**, 284-292 (1994); S.B. Roy, A.K. Pradhan and P. Chaddah, Physica C **253**, 191–193 (1995); M.R. Yoshizaki, J Physical Soc Japan, Vol. **78**, 024703–024708 (2009).
112 R.B. Flippen, T.R. Askew, J.A. Fendrich and C.J. Vanderbeek, Phys. Rev. B **52**, 9882-9885 (1995); M. Tange, H. Ikeda and R. Yoshizaki 1, Phys Rev B **74**, 064514–064522 (2006).
113 C.P. Bean and J.D. Livingston, Phys Rev Lett **12**, 14–16 (1964).
114 L. Burlachkov, M. Konczykowski and Y. Yeshurun, F. Holtzberg, Phys Rev B **45**, 8193–8196 (1992).
115 A.M. Campbell and J.E. Evetts, Ed., Taylor and Francis, London, 1972.
116 M. Konzykowski, M. Burlachkov and L.I. Yeshurun, Phys Rev B **43**, 13707–13710 (1991).
117 L. Burlachkov, M. Konczykowski, Y. Yeshurun and F. Holtzberg, J Appl Phys **70**, 5759–5761 (2001).
118 L. Burlachkov, Y. Yeshurun, M. Konczykowski and F. Holtzberg, Phys Rev B **45**, 8193–8196 (1992).
119 J.R. Clem, in: Low Temperature Physics-LT 13, Vol. 3, Eds. K.D. Timmerhaus, W.J.O' Sullivan and E.F. Hammel, Plenum, New York, 1974.
120 A. Palau, J.L. Macmanus-Driscoll and M.G. Blamire, Supercond Sci Technol **20**, 136–140 (2007).
121 R. Aoki and H.U. Habermeier, Jpn J Appl Phys **26**, 1453–1462 (1987).
122 L.E. Helseth, P.E. Goa, H. Hauglin, M. Baziljevich and T.H. Johansen, Phys Rev. B **65**, 132514–132526 (2002).
123 L.N. Bulaevskii, E.M. Chudnovski and M.P. Maley., Appl Phys Lett **76**, 2594–2596 (2000).
124 A. Hoffmann, P. Prieto and I.K. Schulle, Phys Rev B **61**, 6958–6965 (2000).
125 O.M. Stoll, M.I. Montero, J. Guimpel, J.J. Akerman and I.K. Schuller, Phys Rev B **65**, 104518–104526 (2002).
126 K. Harada, H. Kamimura, T. Kasai, A. Matsuda, A. Tonomura and V.V. Moshalkov, Science **274**, 1167–1170 (1996).
127 D.R. Harshman, G. Aeppli, E.J. Ansaldo, B. Batlogg, J.H. Brewer, J.F. Carolan, R.J. Cava, M. Celio, A.C.D. Chaklader, W.N. Hardy, S.R. Kreitzman, G.M. Luke, D.R. Noake and M. Senba, Phys Rev B **36**, 2386–2389 (1987).
128 M. Cyrot and D. Pavuna, Introduction to Superconductivity and High-T_C Materials, ed., World Scientific, Singapore 1992.
129 G. Papari, F. Carillo, D. Stornaiuolo, L. Longobardi, F. Beltram and F. Tafuri, Supercond Sci Technol, Vol. **25**, 35011–35015 (2012).
130 S. Djordjevic, E. Farber, G. Deutscher, N. Bontemps, O. Durand and J.P. Contour, Eur Phys J B **25**, 407–416 (2002).
131 K. Fossheim, A. Sudbø, Superconductivity, Physics and Applications, John Wiley & Sons Ltd, The Atrium, Southern Gate, Chichester, England, 2004.

132 R. Prozorov, R.W. Giannetta, A. Carrington, P. Fournier, R.L. Greene; P. Guptasarma, D.G. Hinks and A.R. Banks, Appl Phys Lett, Vol. 77, 4202–4204 (2000).

133 D.A. Bonn, R. Liang, T.M. Riseman, D.J. Baar, D.C. Morgan, K. Zhang, P. Dosanjh, T.L. Duty, A. MacFarlane, G.D. Morris, J.H. Brewer, W.N. Hardy, C. Kallin and A.J. Berlinsky, Phys Rev B **47**, 11314–11328 (1993).

134 J.L. Tallon, C. Bernhard, U. Binninger, A. Hofer, G.V.M. Williams, E.J. Ansaldo, J.I. Budnick and Ch. Niedermayer, Phys Rev Lett **74**, 1008–1011 (1995).

135 J.E. Sonier, J.H. Brewer, R.F. Kiefl, D.A. Bonn, S.R. Dunsiger, W.N. Hardy, Ruixing Liang, W.A. MacFarlane, R.I. Miller, T.M. Riseman, D.R. Noakes, C.E. Stronach and M.F. White Jr., Phys Rev Lett **79**, 2875–2878 (1997).

136 V. Lauter-Pasyuk, H.J. Lauter, V.L. Aksenov, E.I. Kornilov, A.V. Petrenko and P. Leidere, Physica B **248**, 166–170 (1998).

137 C. Panagopoulos, J.R. Cooper and T. Xiang, Phys Rev B **57**, 13422–13425 (1998).

138 D.N. Basov, R. Liang, D.A. Bonn, W.N. Hardy, B. Dabrowski, M. Quijada, D.B. Tanner, J.P. Rice, D.M. Ginsberg and T. Timusk, Phys Rev Lett **74**, 598–601 (1995).

Appendice

1. Demagnetizing Fields

1.1. *Origin of Demagnetizing Fields*

In order to understand the origin of demagnetizing fields, let us consider a bar magnetized by a magnetic field applied from left to right, for example. When the applied field is removed, one notes the occurrence of a North Pole at the right end of the bar and a South Pole at its left end. The field lines leave the North Pole and move towards the South Pole inside as well as outside the magnetized bar i.e. in the opposite direction of the previous applied field. This clearly shows that within the material, the magnetic field thus created tends to demagnetize the magnet. This is this action which is the source of the demagnetizing fields. These effects are very important for magnetic measurements and are a constant challenge for theorists of magnetism.[1]

1.2. *Field lines of a Magnet in the Absence of an Applied Field*

The demagnetizing field H_d always acts in the opposite direction to that of the magnetization M which creates it. In the case of a magnet placed in a zero magnetic field, the active field is the demagnetizing field H_d and the relationship between the magnetic induction B which prevails inside the magnet, the applied field H and magnetization M:

$$B = H + 4\pi M \qquad \text{(A.1)}$$

becomes equal to:

$$B = -H_d + 4\pi M \tag{A.2}$$

The induction B within the magnet is smaller than $4\pi M$ but directed in the same direction as the magnetization vector because H_d never exceeds $4\pi M$ in amplitude. Let us note that the lines of induction B are continuous and directed from S to N within the magnet. Outside the magnet, $B = H$ and the external fields are identical. On the one hand, the magnet forms an open circuit because only a part of the flux remains in the body, and on the other hand the magnetization is not uniform there. Indeed, the lines B diverge near the edges indicating that the flux density is weak there compared to that prevailing in the middle of the bar. In other words, the demagnetizing field is very strong near the edges.

Another intuitive definition was suggested by S. Senoussi.[2] It starts from a perfect ferromagnetic system where the domain walls are totally free to move in the presence of an internal field H_{in}. In zero field, these walls are motionless and their distribution is such as the magnetization $M = 0$. The application of a small field induces a movement of these walls until $H_{in} = H + H_d = 0$, with, by definition a demagnetizing factor N_d such as $H_d = -N_d M$.

1.3. Field lines of a Magnet in the Presence of an Applied Field

When the magnetized bar is placed in an area where there is a magnetic field, it deteriorates the shape of the lines of this field. The flux lines are tighter inside the magnet; they are hustled as if they have more facility (more permeability) to pass through the magnet than by the surrounding space. This is the physical origin of the term "permeability μ" encountered in the magnetism. Therefore, in all points outside the magnet, close to its center, the field is reduced. In this case, it is assumed that $\mu > 1$ (case of the iron, for example). The reverse effect occurs for a diamagnetic material where the flux lines try to avoid the body. So, the flux density is larger outside than inside a diamagnetic material.

1.4. Coefficients of Demagnetizing Fields

One can consider the non-uniformity of the flux density B inside the magnet as an escape of the flux lines near the edges. It is possible to eliminate such distortion of the flux lines by carving the sample in cone. One then obtains an ellipsoid with this time a perfectly uniform induction. This uniformity of B results in the uniformity of the demagnetizing field H_d; this last being proportional to the magnetization M that creates it:

$$H_d = N_d \, M \tag{A.3}$$

where N_d is the demagnetizing factor which mainly depends on the shape of the material and which can be exactly calculated only in the case of an ellipsoid. In the particular case of a sphere for example, it is equal to:

$$N_d = \frac{4\pi}{3} \tag{A.4}$$

The general case of an ellipsoid which has three different axes 2a, 2b, 2c, and wherein any section perpendicular to any axis is an ellipse, presents no interest in practice. The most interesting case is rather that of the ellipsoid of revolution also called "spheroid". The prolate spheroid is formed by rotating the ellipse around the 2c-axis. One then obtains $a = b$ and the resulting solid has the shape of a rod. By rotation around the 2a-axis, the flattened spheroid is obtained with $b = c$. It has the shape of a disc. If N_a, N_b and N_c are the particular values of the demagnetizing factor along the **a**, **b** and **c** axes, respectively, these quantities are interconnected in a given ellipsoid by the relationship:

$$N_a + N_b + N_c = 4\pi \tag{A.5}$$

We give hereinafter, without demonstrating, the formulas for calculating these coefficients for the elongate and flattened spheroids.[1]

1.4.1. Elongated Spheroid (rod)

$a = b \neq c$ and by posing $r = c/a$, one gets:

$$N_c = \frac{4\pi}{r^2 - 1} \left[\frac{r}{\sqrt{r^2 - 1}} \ln(r + \sqrt{r^2 - 1}) - 1 \right] \tag{A.6}$$

$$N_a = N_b = \frac{4\pi - N_c}{2} \tag{A.7}$$

When r is very large (very large rod), the following equations are obtained:

$$N_a = N_b \sim 2\pi \tag{A.8}$$

$$N_c \sim \frac{4\pi}{r^2} (\ln 2r - 1) \tag{A.9}$$

It is seen that N_c approaches zero when r approaches infinity (example: for $r = 10$, $N_a = N_b = 6.15$ and $N_c = 0.256$).

1.4.2. Flattened Spheroid (disc)

Knowing that $a \neq b = c$ and $r = c/a$, it comes:

$$N_a = \frac{4\pi r^2}{r^2 - 1}\left(1 - \sqrt{\frac{1}{r^2 - 1}} \sin^{-1}\frac{\sqrt{r^2 - 1}}{r}\right) \tag{A.10}$$

$$N_b = N_c = \frac{4\pi - N_a}{2} \tag{A.11}$$

When r is very large (very thin disk), one has:

$$N_a \sim 4\pi \tag{A.12}$$

$$N_b = N_c \sim \frac{\pi^2}{r} \tag{A.13}$$

N_b and N_c approach zero when r approaches infinity (example: $r = 10$, one has $N_a = 10.82 = N_b$ and $N_c = 0.88$). In practice, the most used samples are the magnetized cylindrical rod along its axis and the magnetized cylindrical disc in the disc plane. These samples are never uniformly magnetized except when they are fully saturated. Strictly speaking, the demagnetizing field varies from one point to another in the sample and cannot be accurately calculated. The experiment shows that the demagnetizing coefficient depends on both the shape of the sample and its permeability. Finally, let us note that in the SI system where B, H and M are related by the equation $B = \mu_0 (H + M)$, the demagnetizing factors are $1/4\pi$ times the values in the c.g.s. system. Thus, in the case of a sphere, $N_d = 1/3$ and $N_a + N_b + N_c = 1$ for an ellipsoid are obtained.[1]

2. Physical Quantities in SI and C.G.S. Gauss Systems

For clarity, we have used in this book depending on the situation three unit systems, the international system (SI), the c.g.s.-gauss units system and the convenient c.g.s. system. In the convenient c.g.s. system, the magnetization per unit volume is expressed in emu/cm³, the magnetic field in Oersted (Oe) and the current density in A/cm². These units are derived from the c.g.s. gauss units by simply replacing the speed of light (c) by 10. To help the reader in search of more uniformity, we give below a brief correspondence between the SI system and c.g.s. gauss units and the expressions of some useful equations written in the two systems. An extensive review of these unit systems can be found in the references 3 and 4.

Quantity	S.I.	c.g.s. gauss
length	meter (m)	1 cm = 10^{-2} m
Mass	kilogram (kg)	1 g = 10^{-3} kg
Time	second (s)	1 s = 1 s
Force	newton (N)	1 dyne = 10^{-5} N
Work, Energy	joule (J)	1 erg = 10^{-7} J
Power	watt (W)	1erg/s = 10^{-7} W
Quantity of electricity	coulomb	1 emu = 10 C
Amperage	ampere (A)	1 biot = 10 A
Voluminal density of current	ampere per square meter (A/m²)	1 biot/cm² = 10^5 A/m²
Electrical field	volt per meter (V/m)	1 esu = 3 × 10^4 V/m
Resistance	ohm (W)	1 esu = 9 × 10^{11} W
Magnetic flux	weber, maxwell	1 Maxwell = 10^{-8} Wb
Magnetic induction (B)	tesla	1 gauss = 10^{-4} T
Magnetic field (H)	ampere per meter (A/m)	1 oersted = $\dfrac{10^3}{4\pi}$ A/m
Magnetization per unit volume or magnetic intensity (M)	ampere per meter	1 emu/cm³ = 10^3 A/m
Magnetic moment	A.m²	1 emu = 10^{-3} A.m²
Inductance	henry (H)	1 emu = 10^{-9} H
Bean formula for a cylinder	$J = \dfrac{3\,M}{R}$	$J = \dfrac{3c\,M}{R}$

3. Some Useful Equations

SI	c.g.s. gauss	Comment
$\mu_0 = 4\pi\ 10^{-7}\ (H/m)$	1	μ_0 is the permeability of the vacuum.
$d\mathbf{B} = \dfrac{\mu_0}{4\pi}\ I\ d\mathbf{l} \times \dfrac{\mathbf{r}}{r^3}$	$d\mathbf{B} = I\ d\mathbf{l} \times \dfrac{\mathbf{r}}{r^3}$	**B** is the magnetic field.
$\oint \mathbf{H}.\,d\mathbf{l} = NI$	$\oint \mathbf{H}.\,d\mathbf{l} = 4\pi\ NI$	**H** is the excitation field.
$\mathbf{J} = \sigma\ \mathbf{E}$		σ is the conductivity of the material.
$\mathbf{M} = \dfrac{\mathbf{B}}{\mu_0} - \mathbf{H}$	$\mathbf{M} = \dfrac{B - H}{4\pi}$	**M** is the voluminal magnetization.
$d\mathbf{m} = I\ d\mathbf{S}$	$d\mathbf{m} = \dfrac{1}{c}\ d\mathbf{S}$	dm is the magnetic moment of a small circuit crossed by a current *I*.
$\mathbf{F} = e\mathbf{v} \times \mathbf{B} + e\mathbf{E}$	$\mathbf{F} = \dfrac{e}{c}\ \mathbf{v} \times \mathbf{B} + e\mathbf{E}$	**F** is the force acting on a particle of charge e, driven with a speed *v* and subjected to the electrical field *E* and the magnetic induction *B*.
$\text{rot } \mathbf{H} = \mathbf{J} + \dfrac{\partial \mathbf{D}}{\partial t}$	$\text{rot } \mathbf{H} = \dfrac{1}{c}\left(4\pi\mathbf{J} + \dfrac{\partial \mathbf{D}}{\partial t}\right)$	**D** is the electrical induction. The associated current is negligible in the experimental conditions considered in this book.
$\mathbf{E} = -\ \nabla V - \dfrac{\partial \mathbf{A}}{\partial t}$	$\mathbf{E} = -\ \nabla V - \dfrac{1}{c}\ \dfrac{\partial \mathbf{A}}{\partial t}$	**A** is the vector of the electrical potential and *V* the electrical potential.
$\mathbf{B} = \text{rot } \mathbf{A}$	$\mathbf{B} = \text{rot } \mathbf{A}$	This equation is identical in the two systems of units.

REFERENCES

1. H. Zjilstra, Experimental Methods in Magnetism, Wiley, New York, 1967.
2. S. Senoussi, personal communication (2003).
3. J.D. Jackson, Classical Electrodynamics, 2nd edition, 1962, 1975, by Wiley & Sons.
4. A.S. Arrot, Ultrathin Magnetic Structures, Eds. J.A.C. Bland and B. Heinrich (Springer-Verlag Berlin Heidelberg, 1994, p. 7).

Index

Printed and bound by CPI Group (UK) Ltd, Croydon, CR0 4YY

01/11/2024

01782622-0018